ABSOLUTE

ABSOLUTE

DEJAN STOJANOVIĆ

NAB

New Avenue Books

New Avenue Books

Second Edition

Library of Congress Control Number: 2024953156

ISBN-13: 978-1-966571-09-4

TABLE OF CONTENTS

PREFACE

> "In the eighteenth century, philosophers considered the whole of human knowledge, including science, their field and discussed questions such as: Did the universe have a beginning? However, in the nineteenth and twentieth centuries, science became too technical and mathematical for the philosophers, or anyone else except a few specialists. Philosophers reduced the scope of their inquiries so much that Wittgenstein, the most famous philosopher of this century, said, "The sole remaining task for philosophy is the analysis of language." What a comedown from the great tradition of philosophy from Aristotle to Kant."[1]
>
> — Stephen Hawking, *The Theory of Everything*

Indeed, many vital questions would have remained unanswered and unresolved without philosophy. The same applies to the future of humankind and its destiny. From childhood, we all start asking questions about the world around us. At age ten, I realized an atom is a microsolar system. This idea became almost an obsession in my head, which led to constant pondering and philosophical considerations. Since then, many other thoughts have preoccupied me, some of which I chose to present in this book.

Between the ages of 19 and 26, I put on paper many ideas and wrote essays, not only philosophical but also concerning literature and language. These manuscripts disappeared after the unfortunate events in Serbia and Kosovo in 1999[2]. Since it was impossible to

---

[1] Hawking, S. (2003). *The Illustrated Theory of Everything: The Origin and Fate of the Universe* (p. 109). Beverly Hills: Phoeinix Books, Inc. (2003).

[2] Bombing of Serbia in 1999.

recover the lost manuscripts, I planned to write down the old ideas around 2000-2001, but some business activities did not allow that. Finally, I wrote down the old ideas and worked on some new ones between 2007 and 2012. Due to other obligations, I could not complete the minor task of adding final touch-ups and finishing the book until recently. The book contains the main ideas I have been dealing with since I was 18.

My original goal was to write a book no longer than 100 pages without overwhelming the reader with too many details. I intended to thoroughly present some ideas and hint at others using as few words as possible. Although I am unsure if I achieved the desired simplicity in the book, I was aiming at it. The beauty itself is pure and simple. It is enough to cast our eyes around and notice the miracles of simplicity. Simplicity also functions in nature, the arts and sciences, or anything we do. The truth, like beauty, should also be simple to exude its uttermost meaning and purpose.

The world is simultaneously real and beautiful, regardless of the suffering that occurs almost everywhere to a lesser or higher degree. Beauty is the glorification of the world and existence at their best. By praising beauty, we improve, among other things, our constant development and evolution in search of meaning. Just as beauty shines in the world, we can recognize the brilliance of the esthetics of science and philosophy in the most significant scientific achievements.

Without relationships and our senses, nothing "exists," including the experience of beauty. The world would "not exist," as we see it, without our eyes and senses. Even if myriad beings existed without senses, they would all be mute. Senses offer sense and meaning to the multitude in the labyrinth of the Universe. Thus, the underlying meaning of this book is the meaning itself.

D. S.

CHAPTER 1

THE RELATIONSHIP BETWEEN THE ESSENCE AND FORM

Nothing exists in the way we see it except senses, thoughts, and ideas that are able and capable of experiencing the world. This idea was already established in ancient Greece by Democritus, some 2,000 years before Descartes, based on his theory of conventions. After an extended period, other philosophers and scientists, starting with the first modern philosopher, Descartes, treated this subject thoroughly and systematically.

There is no beauty without eyes to enjoy it; likewise, there is no ugliness and deformity. The most beautiful fragrance does not exist without a sense of smell. There are no borders without having the sight to establish them; there is no beginning and end without "time." Only we pass through time, measuring our growth and observing our disappearance. Time does not pass. Time does not measure us; instead, we measure time and are the measure of time.

There is no time without space or space without the division of the Being, oneness, to establish distances among the shapes. If we remove the shapes from space, not only do the shapes disappear, but space disappears. What is left is emptiness (void); the absolute vacuum is not space but Nothing. Shapes (forms, objects) and their dispersion create the "illusion" of space. In every form (object), there is emptiness. Owing to emptiness, every object is a universe, of and for itself, which contains other universes within itself.

Emptiness inhabits the world and "lives" within objects. Without emptiness, there are no dimensions, distances, or lengths; without an extent, there is no distance, path, motion, or relationships. Without the world, emptiness or nothingness is the "finite infinity." It is everywhere and nowhere at the same time. Every point is the same point.

Paradoxically, space creates time. These are not two categories; one cannot exist without the other. Something that does not exist—emptiness or nothing—is the only possible space. How is it possible that Nothing becomes Something and that it creates Something, renders it possible, and gives it life? The Absolute One makes it possible to convert and divide itself into a plurality by moving emptiness into itself, the Being. By swallowing emptiness, parts make a plurality of each part. All we consider or see as space is only the emptiness of the divided Absolute. Yet this is only a converted manifestation of the Absolute, which always stays the One. The original state and oneness of the Absolute are never lost.

Absolute is infinitesimal, almost nonexistent "materially." It is smaller than any conceivable shape or object, smaller than a quark, not even the size of the Planck length. In its volume, every divided world and Everything inhabited by emptiness is "bigger" than the Absolute. But only this nothing, emptiness, makes the life of the Absolute possible. If the One is undivided, it is without motion; the power becomes a weakness because it is without an exit.

The exit of the Being is not in exiting itself but instead leaving on a voyage. Since there is nothing except the Being and Nonbeing (the Nothing, emptiness), it cannot go on a journey because a journey into the void has no meaning or purpose. Alone within itself, at the height of its power, it becomes the same as its opposite—Nothingness. The Being resolves this problem not by moving into emptiness but by moving into itself, by sucking up nothingness, or void, into itself.

The passing of the Absolute through itself, between two zeros or two nothings, is life. This passage, which makes life possible, is impossible without nothingness as an equally important "force." Only nothingness is emptiness or an absolute vacuum. The Absolute Source creates the world and life by swallowing emptiness, not by the explosion into the void. The Big Bang is not the explosion of the Absolute Source into emptiness but rather the silent explosion of emptiness into the Absolute Source. Big Bang is the wormhole of existence, connecting the Source beyond the spacetime continuum to its creation, the World. The power of the Source creates and materializes itself through the "Gate" of Zero. Zero is the passage, or wormhole, between Nothing and Everything. There is no energy more enormous than this one. If we imagine Hell as a space of vast forces, then there is no better place for it than in the world.

ORDER

The main characteristic of the Being, or the Universal Mind, is its creative power, which manifests itself superbly in the creation of the Universe. The compressed idea of the Universal Mind is dispersed into the world by the "explosion"—the Big Bang. The creation of order is not a compression of energy but the ordering of elements. Einstein treats the effect, not the cause. Still, we need to explain, at least approximately, the cause and the most relevant relationship—the relationship between the Being and the Nonbeing or Nothingness.

Our primary interest is the Absolute of the God-World. The world is relative, and Everything in it is relative, but before the effect of the world, as we know it, the world is absolute in its primordial stage. This Absolute World, or Universal Mind, also has the potential to create an infinite number of worlds or universes. Since creation is not independent of its creator and is the creator itself in a different form, the creation process is the process of recreation. The only Absolute is the Absolute of the unborn world when the Absolute of Nothingness is Everything. The ultimate Theory of Relativity shall explain the relationship between the Being and the Nonbeing or Nothingness. This relationship is the "place" where the main game takes place. For thousands of years, ideas about relativity and the strange features of time have existed in literature.

The modern atom is not the one that Democritus imagined. It only owes its name to Democritus but does not represent the smallest particle. Democritus had precisely the smallest particle in mind—a quark, string, or something smaller up to the "zero" point of Everything. Science needs to understand the smallest uncuttable (indivisible) particle.

Small particles succeed in uniting, and their ordering is not regulated by smaller or bigger masses because, in the primordial state, there are no bigger or smaller masses. The mover puts everything in motion or creates an illusion of movement. The mover produced enough vibration to cause the formation of quarks, electrons, and possibly "other particles" from the primordial foggy fluid and fire in trillions of degrees.

The first spark puts Everything in motion, sufficient that Everything seeks its place unmistakably based on the principle of chance or a program. Evolution makes no difference, whether chance or a program, because Everything will play out according to "plan" and establish order. If we suppose that the whole future development lies in the principle of chance, then this chance must be so perfect that it is irrelevant if the program develops perchance or under the supervision of a higher power. The result would be the same. (The chance offers a bigger chance, for in endless cycles of the Universe and life, the chance always empowers or enables a new world and order.)

Everything takes place based on the principle of motion; motion establishes the law, not the energy or mass. The creation of more significant masses is the consequence of motion, not its cause. There are only bigger and smaller families; bigger ones only take up more space, but all families are the same.

In diluted reality, the whole world is the same; the micro-dimension is the same everywhere; the macro dimension is the illusion of the senses and eyes. Macro-shapes exist, but only as of the sum, not as inherently different. The basis of every star and every galaxy is the same—antimatter or dark matter and dark energy hide similar or opposite laws in themselves. However, although they are similar or opposite to the visible ones, they are based on the same laws of the bigger or smaller "particles," shapes, and masses. The same principles are everywhere, from the visible to the invisible world or in the world incomprehensible or beyond the cognitive ability of humans.

The world itself is relative, but what about the absolute in the relative? What can we say about absolute relativity or the relativity of the Absolute? Can the Absolute be relative? Can relativity be absolute? Where is the point of reconciliation? Is there such a point?

ABSOLUTE

The idea of the Absolute implies perfection. Absolute is impossible if it is not perfect. Nevertheless, is it possible to conceive anything perfect? Everything that exists has shortcomings. Even if we imagine something as perfect, it cannot be absolute perfection. It does not capture the whole and can only exist in relation to other phenomena. The world is exposed to change at any moment and does not function as an absolute perfection but as a flaw, evolution, and development.

What, then, can be the Absolute? The most logical answer is that nothing can be absolute. This answer hides a few paradoxes. If nothing can be Absolute, then there is no absolute. It is impossible to conceive something absolute in the smallest and the biggest, functioning perfectly without the possibility of a mistake or chance. Impossibility of this kind leads to the erosion of the perfection of the Absolute.

What can be conceived as the Absolute if nothing can be absolute? We can imagine the Nothing as the Absolute. Nothing(ness) can be the Absolute is the most logical answer. This Nothing is perfect. Different meanings of the same words prove the nature of relativity and the relativity of language. Nothing can be the Absolute, but only the Nothing can be the Absolute. Moreover, we do not refer to the nuances of two different meanings here but rather opposed meanings and ideas. It seems that a paradox, sometimes, can be a shortcut to the truth.

Since it is possible to conceive a paradox in almost Everything, it is possible to imagine the truth as a paradox. If it is possible to conceive the paradox in the essence of truth, then it is possible to conceive the truth as the most seductive and deceiving power. Everything singular can be conceived and, sooner or later, comprehended by thought. Everything that is not true has the potential to be disclosed as such—not accurate. However, in its totality, the final, absolute truth is beyond reach—it is hiding and shifting. Still, as perfection, the Absolute cannot be untrue because, in that case, it would not be perfection. Therefore, it must be perfect and true at the same time. It is possible to conceive the Absolute as the truth. In this sense, it is possible to conceive that Nothing is the truth.

Throughout history, philosophers and religious thinkers have dealt with the question of God and Being and Nonbeing or Nothingness. We may think about God as something absolute that nothing bigger is possible to conceive, following St. Anselm's or other ontological arguments from Aquinas to Descartes and others. If we put aside all the objections to the ontological arguments, the main question remains: How did that something, being the biggest or smallest, come into existence? This way, we move into the absurd territory of "infinite regress." This paradox is possible to resolve by getting rid of paradigms and by accepting the fact that the Nothing, regardless of being nothing, has an equally important role or a "dimension," conditionally speaking, as Something has, even if it is the Biggest, Absolute Being, or God. Nothing is also uncreated, unborn; it always was.

Just of itself—nothing is nothing, but just of itself, something, even the biggest something or God, is nothing too. Nothing is uncreated and doesn't need a creator. However, the world would not be possible without this uncreated Nothing, regardless of God's "omnipotence." Since God cannot create nothingness, its omnipotence is contingent upon nothingness. If omnipotence is contingent upon anything, it is not absolute omnipotence, even if this anything is nothing.

If the Being or Nonbeing is uncreated, they both "exist," but the world, as we perceive and experience it, cannot exist without both. Although Nonbeing is nonexistent, its nonexistence is the basis of the world's existence. Only the Absolute encompasses both Being and Nonbeing. The omnipotence of the Being, God, cannot be exercised without the Nonbeing, which means that this omnipotence is not absolute. Nevertheless, as the sum of the Being and Nonbeing, the Absolute is not contingent upon anything outside itself.

Life and the evolution of the world are possible only in the interaction between the Nonbeing (Nothing) and the Being (God). However, what is the real Something, beyond the world, above Something and Nothing or Being and Nonbeing? What is the "force" that unites the world and enables creation, evolution, and life? That Ultimate Force is the Absolute. Only the Absolute pervades Everything.

It is possible to conceive the world even before its "becoming" in one form or another. It is possible to imagine nothingness as always being in the same mode of nothingness. Therefore, the terms we use interchangeably, the first essence of the World, the Universe, or God, do not have a meaning without the opposite "essence," which is without any essence, and that is—nothingness (void, emptiness). Only the "collision" or merger of these two factors—of the Being and Nonbeing—can establish the World as we know it and make it function.

The world without the world (in it) can be God, but this is a God that sleeps. Emptiness, without God, is unfertilized nothing. The Universe is born in the "collision" or merger of an awakened God or Being and Nonbeing or nothingness. Furthermore, the Absolute is "above" emptiness and God. Absolute unites God with Nothingness; absolute enables life. Beyond it, there is nothing; nothing enters and comes out of it. Absolute is above the Being

(God) and Nothingness. Being and the Nonbeing are the Absolute's two sides (latent "forces").

The idea that the World can create itself from nothing is equally absurd as the idea that God created the World from nothing. The lack of the material world, space, and time in the primordial, immaterial state of the Being is not proof that the primordial Being does not exist. Also, the lack of a material world in the primordial state does not mean that the Being creates from nothing. The Creation, or the process of creating, happens in the interaction between the Being and Nonbeing. The immateriality of the primordial Being does not mean that there was only absolute Nothing in the imagined "beginning." This Immaterial Something is the Source of Everything.

The lack of understanding and some consensus about these matters shows the fanaticism of both religious people and atheists who have become atheists more because of hate toward religion than because of "hatred" toward God. Therefore, they cannot understand the difference between God and religion. Curt Gödel[3] (1906—1978) noted, "[He is] against religions but not against religion." Without intermediaries and "holy books," real religion is the desire to reconnect (*religio*) to the Ultimate Source.

The world from nothing is impossible, and life from nonlife is impossible. The world was always here, in one way or another, and nothingness was always here. The World is the Universal Mind's "program" of the Absolute that, by involving nothingness, becomes the Universe. However, the Universe still bears all that is known and unknown to us with the spark of life in it.

[3] Logician, mathematician, and philosopher Gödel was one of the most significant logicians in history and a close friend to Einstein, who enjoyed long walks and conversations with him.

THEORY OF THE ABSOLUTE

The smallest hides the law of the biggest. If it is “separated,” the smallest is the “absolute” of its Universe, although nothing is separate from the Absolute. Absolute remains whole when it dissolves into a plurality or compresses into oneness or singularity.

Greek atomists Leucippus, his pupil Democritus, and other metaphysicians knew that the world was not how people saw and perceived it. If they understood atoms even then, the indivisible particles, not modern new atoms, we must believe that they understood much more. If they knew that every sense and sensation is a convention, they could have understood that space and time are conventions, too. Regardless of not thoroughly understanding or elaborating on these concepts, they understood that the world must be something different from what is experienced by the senses or how the senses understand it. If Everything is by a convention of senses, then senses can represent things differently; that is why the eye watches, not the ear. If Everything is a convention, then Everything we experience by senses must be relative.

Maybe Democritus did not understand that matter, as a convention or the world of atoms that makes the whole Universe, including our brain, which “rules” the senses and cognition, is the same convention. However, an atom is a convention. The World is a Convention of the Absolute. The conventions must be relative; only the Absolute is unconventional, but only by and through conventions are the world and life possible.

The Absolute is "finite" but infinite in its potential. This statement is only theoretical because there is no finitude or infinity in the Absolute without the world as we see it, for the Absolute is limitless beyond the spatial and temporal realm. We can think of limitlessness, infinity, and finitude only in the context of their potential.

The end is only possible if there is a beginning.
Something with the beginning has an end.
Nonbeing has no beginning.
The Being, the God, has no beginning.
Nonbeing is absolute in its nothingness.
Being is absolute in its somethingness.
Nonbeing and Being, if separated from one another, are the same.
Absoluteness makes nothing and something dead or asleep.
Absolute contains all Being and Nonbeing.
Being and Nonbeing, isolated from one another and independent, both become absolute and, for that reason, must annihilate each other. Without this relationship, there is no life.
Without life, there is only absolute nothingness. This Absolute is the only true Absolute.
Relativity is possible only in life, the Universe, and through the world or life.
Without relativity, God is dead.
The savior of the Absolute is relativity itself.
Perfection is a dead end.
The basis of every creation is purpose and meaning.
Absolute cannot have a purpose and meaning without relativity.
Relativity means plurality.
Only in the plurality of the world life is possible.
Absolute singularity is equal to Zero or nothingness.
The power of the Absolute is in its absolute potential for relativity.
Absolute is an engine—*Perpetuum mobile.*
Only Nothing can be Absolute.

Zero of God is still Zero. ∞
Zero of Nothingness is still Zero. ∞
Zero plus Zero is still Zero.
$A = 0 + 0$
Absolute is the sum of two zeros.
$0 + 0 = 0$
Absolute is an ultimate Zero.
$A = B + N$ or ($A = S + N$) (where A is Absolute, B is Being, and N is Nonbeing, or S is Something and N is Nothing)
Infinity is either the absolute finitude or absolute potential.
$\infty = 0$

RELATIVITY OF THE ABSOLUTE

Although the idea of the Absolute's relativity may seem absurd, the absoluteness of the Absolute reduces it to nothingness. Absolute gains life by empowering its components, the Being and the Nonbeing. Being and Nonbeing are the poles of the Absolute. Nevertheless, the Absolute is dead if there is no relation between the Being and Nonbeing. The absolute world is frozen. At the level of a complete absolute, Being and Nonbeing are the same. There must be a love-making between the Being and Nonbeing for the world to exist and for the Absolute to gain, regain, and maintain its existence outside and beyond its absolute form.

Absolute is complete. As something complete, it is also wholly finite. The finitude of the Absolute comes from its completeness. The completeness of the Absolute comes from the Being and its potential infinity. Only the Absolute can be complete because nothingness has no properties to be complete or incomplete. However, the completeness of the Absolute is the source of its demise. It is the "finite" world of infinity. It is the infinity of the finite. It is endlessness with an end. It is the end without endlessness. It is the absolute and "omnipotent" zero.

At its absolute point, the Absolute is equal to itself and its components: There is no difference between Being and Nonbeing. Since there is no difference between Being and Nonbeing, there is only nothingness. The absoluteness of the Being reduces it to its opposite—the Nonbeing. The absoluteness of the absolute becomes the absoluteness of the Nonbeing or Nothingness. The world of a complete Absolute is sterile.

Beyond all paradoxes, the purpose of the World is to be alive and live. To live, it must give birth to relations. Relations are the source of relativity and life. To exist, the poles of the Absolute must be active and interact with each other.

The first and most important relationship is the Being and the Nonbeing relationship.
This relationship enables the world to put itself into motion.
Existence is the source of relativity and relations.
Relations are the source of life.
Relativity is the source of life of the Absolute Being and the Absolute Nonbeing.
Once in a relationship, the Absolute Being becomes a Relative Being, and the Absolute Nonbeing becomes a Relative Nonbeing.
The relativity of the poles of the Absolute creates the relativity of the Absolute.
Without relativity, absolute would not be possible.
Without an absolute, relativity would be meaningless and accidental.

ABSOLUTENESS OF RELATIVITY

The ultimate paradox is how the relative can simultaneously be absolute. The truth itself is a paradox. When we think about the forces of the Universe, we always think about physical forces that we are familiar with, mainly through science. Nevertheless, the main forces from which Everything comes into existence are different.

What is a force? Must it always be active and exercise some impact on something else, or can it be passive, without any measurable effect, and still be an equally essential and potent force? The answer to this question is—yes. The Absolute's two main forces are the Being and the Nonbeing (nothingness). Without one, the other is dead. Without nonexistence, there is no existence. Nothingness or the Nonbeing represents nonexistence, but without this Nonbeing or nothingness or nonexistence, the Being has no chance. Without nothing, something is nothing, too. Since the two main players in the game of the Absolute cannot exist without one another, it means that they are relative.

Conversely, since there is absolute quality in their interdependence, this relativity is absolute. The main rules of the game always stay the same. The Being will always be the Being; the Nothing will always be the Nothing.

What makes them most absolute is their separateness. "Persisting" in separateness, not only do Being and Nonbeing resemble one another, but they cancel every difference between themselves. They become one. By becoming one, they become the

Absolute itself, at which point the Nonbeing rules. Everything transforms into nothingness.

There is an absolute quality to the Being, Nonbeing, and the Absolute. The Absolute is neither the Being of itself nor the Nonbeing of itself but rather a sum of them. Absolute (conceptually) unites the Being and the Nonbeing. As a result of the two, Absolute defies its absoluteness, for it would be logical that the other two result from it. Its existence depends on them. Without the relative, the Absolute becomes not only relative but irrelevant and impossible. We call this the relativity of absoluteness. Paradoxically, the relativity of absoluteness is the source and the creator of the absoluteness of relativity or existence.

INFINITY AND FINITUDE

Any dimension or number is a relative category. The number is absolute in its absolute volume, but it has its absolute volume only as a sign. No number is absolute; every number gains meaning in relations and only exists in relationships. Only the infinity is absolute and "contains every number" (potentially), but there is no last number. Infinity, as a manifestation in any of its variations, possibilities, and cycles, cannot be reached because infinity is an absolute potential. It is not possible to reach the last number. If there were the last number, there would be no infinity. Although there can be no infinite regress, there can be infinite "progress." The idea of the infinite regress is the intellectual construct or attempt to reconstruct the "past conceptually." In contrast, infinite progress is potential that may actualize up to every actualized point *ad infinitum*. There is no end to the potential of the Absolute.

Everything that exists is finite. The only infinity is the world's potential for infinite variations. Absolute void, or emptiness, is infinite from the world's perspective. Still, from the perspective of itself, without the world in it, the absolute void is the same everywhere, which means that every point is the same point, and this means that it is finite; it is simply nothing. Also, the void itself represents nonexistence. Before its actualization as a plurality, the world is finite and absolute in its singularity. It becomes equal to its antipode—absolute void or emptiness—which makes it nonexistent.

Infinity is a contradiction in itself. The only way for the concept of infinity to function appropriately in the world is to be imagined because if there is infinity, there is no end, which is impossible. Anything that exists must be "finite," although there is no end to the world's potential and its cycles. In this sense, absolute is infinite. Infinity is the potential, whereas concrete manifestations of the Absolute are finite. Any particular Universe is the manifestation of the Absolute (Universal Mind) and is finite. For that reason, universes, creation, and life are not repetitions of the absolute potential but are representations of this potential in an always new yet finite way (which does not exclude potential recurrences). The Universal Mind must be infinite in a way that the Nothing is infinite—everywhere and nowhere. In that state, everywhere is nowhere, and this nowhere is everywhere.

Infinity, strictly speaking, is not compatible with existence or life. Still, finitude and infinity are compatible. Their compatibility lies in the infinite potential of the Being. Although Being is finite in any actual realization, its potential and existence are infinite in a broader sense. The main savior of existence is not the world as a repetition of similar or the same cycles, but a continuation of always new births, rebirths of Universes, not the copies (although they are possible) of the previous ones but always new and different, based on the absolute potential of the Being for infinite variety.

Possibilities are infinite. Absolute is infinite in its potential and unaware of all possibilities. Every new cycle explores the new possibility of where free will lies. The Universe is "learning" its potential and creative power from itself through its endless cycles and variations. Its cycles are not the same. Here lies the spark of freedom and potential for free will that makes determinism and free will compatible.

Nothing represented by a number is absolute; every number is divisible. When divided, it stays the same in the volume of its limited absolute, which lives in its relative volumes. Even one can be divided into "infinity"; not only the Absolute or the Absolute One, but a "small," "regular" one can be divided within itself almost without a limit.

The point where the One cannot be further divided is the gateway to infinity, which is Zero. If we imagine this point is the Plank length, the "point" beyond it is the spotless spot or pointless point of the Being, immaterial and Zero. The Big Bang is the whisper of awakening eternity, awakening ultimate Being, essence. The Big Bang is the ultimate kiss between the Being and the Nonbeing, lovemaking from which their child, the Universe, is born.

The voyage toward infinity is a voyage without an end. The biggest is not closer to infinity than the smallest. The voyage toward infinity is the voyage through the biggest and the smallest. The biggest is closer to one imagined "edge" of infinity, and the smallest is closer to the "other." At this point, they meet. However, infinity is only one.

We cannot seek infinity only from one side or from one point; the way toward infinity is in both directions. The biggest lives in the smallest, and the smallest lives in the biggest; there is only one organism. The smallest and the biggest lead to the same place, to the meeting point of the biggest and the smallest.

Infinity, or absolute multitude, is impossible based on logic or numbers. If there were no possibility for particularities (any particular entities or worlds) to end, that would be the end of the world; the world would transform itself into a chaotic assembly of numbers, shapes, and not life. Luckily, the world is one, and that One is absolute with its value and "volume." This "volume" can dissolve within itself but is impossible without the "actual infinity" (potential).

(Universe is one. All potential or existing "parallel universes," whatever we call them, are part of the same Universe. If the Universe is not one, then it is not a real universe, but only our idea or the name applied to something we chose to call the Universe.)

The actual "infinity" is nothingness, yet nothingness, in its primordial stage, undisturbed by the Being, is nonexistent, which means finite, except as a potential for space and infinity. In its primordial stage, the Being is complete, which completeness makes it finite, except as a potential for infinity. Absolute completeness of the Being in its original stage transforms it into nothingness, and they become almost the same.

The only savior of the Being is its creative power. The power to create or recreate itself in countless forms is the source of life and the reality we call the Universe, but it is also the source of life for the Creator itself. Without its creative power to transform, the Being would become frozen, almost dead. Absolute, in its primordial form, is complete and therefore finite, yet infinite as a potential. Without multitude, creation, and motion, the absolute potential is asleep. The absolute potential is almost useless and dead if not transformed into plurality. The Universe is the Child of the Being and the Nonbeing. This Child is at the same time its own Father and Mother.

Only nothingness has no beginning or end. The One does not have a temporal beginning or end, but the One has the capability or power of dissolution or dividing. Zero is absolute: it cannot be divided by itself or others or subtracted or deducted. Only the Zero can capture or fill in the whole "infinity" because it does not need a new number and then a new number and a new one again, which is absurd because the number itself would, in that way, sink and die of itself lost in its infinity.

Although we may disagree with the Aristotelian idea that we can figure out the laws of nature by pure thought, we still believe that strict reliance on experiments and observation is insufficient to answer the most profound questions of science and the Universe. For this reason, science has stagnated regarding these questions.

Infinity is a concept and not necessarily a reality. If infinity existed in reality, all that exists would have to be infinite in all directions on a macro and micro level. There is no infinity if there is an end in any direction or level, micro or macro. Infinity would mean endless cycles of matter and universes on every side. Infinity, in actuality, is illogical and impossible, and it defies what we already know about the Universe.

Limits provide the means to form. Without limits, there are no forms; the universe loses its meaning. On the other hand, limitless cannot have a form and cannot be formed; it is chaotic and meaningless.

We mostly envision infinity as something beyond our reach and in outer space. However, if it existed, infinity would have to expand on a micro-level within something we may imagine as the inner space. If we say it this way, infinity cannot be only outside; it must also be "inside" with no end to the material universe anywhere in outer space and on its micro-level. It also means that, if we simplify the idea, the tiniest particles, quarks, "strings," or waves, would be larger, compared to what they contain within themselves, than the visible Universe in comparison to us.

The only "thing" without any properties is absolute void, without the Universe in it. This void, or absolute vacuum, has no beginning, end, or time; therefore, it lacks limits and limitations. This featureless void is more limited than anything we can imagine. It is the same everywhere; it is not even infinitesimal. It is nonexistent, yet nonexistence is the greatest limit of all.

Something that is formed or exists must have limits. Paradoxically, these limits are the source of infinity as a potential, not an actuality. The actuality of infinity is never-ending existence and life, and not a literal infinity *per se.* Infinity represents the infinity of possible variations of that which is inherently finite. Finitude is, therefore, not incompatible with infinitude.

At the speed of light, the mass unites every particle (Everything) in a unified "vibration." At the speed of light, the "kinetic" force would be equivalent to the force of infinite mass. It is not that the mass becomes infinite, but the force of "mass," "gravitational" pull, as a consequence of speed (motion), becomes as strong as if it were infinite. The source of gravitation is not mass but motion and speed (kinetic property of mass and matter). This "point" is the door between our world, as we experience it, and the primordial world. This "point" is the gateway between noumenon and appearance, between the noumenon and phenomenon.

U=I+M (UNIVERSE = INFORMATION + MOTION)
Creation = Explosion of Information
Disappearance = Absolute speed or the lack of motion = Total stop

At the point of "absolute speed," the mass, energy, or anything we consider material disappears. Absolute speed is "speedless." Absolute speed is always at the same point, which is the pointless point: speed without the speed, the point without the point—beginning, at the same time as the end. Absolute speed is motionless. Without motion or speed, there is no mass. Without motion, the mass disappears. Without motion, all laws of physics stop functioning.

Motion is the unifying force of Everything. Without motion, no other forces exist on the micro or macro level, quantum physics, or the Universe level. Motion is the force unifying the whole Universe in one Unified Field. This Unified Field is one organism, and we call it the Universe. Without motion, there is no unity. What powers the motion is the Universal Mind. The Universe is a Child of the Universal Mind. Although the Universe is a "program" of a Universal Mind, there is still freedom and free will based on the potential of the Absolute to always be different in its infinite manifestations.

The last division of the indivisible, which cannot be divided anymore, is the "moment" of disappearance. "Universal Mind" creates energy and mass through the universal programming of Reality. This Reality has a beginning and end: it is not eternal and can disappear. Matter is the program of the "Universal Mind." The "point" of something we call the Big Bang is the "point "of creation or recreation. Everything that disappeared is created or recreated again (not as an ordinary, identical recurrence, although even that is possible in infinity).

New life is not a repetition of a previous one but an entirely new circle. Theoretically, repetitions are possible because there is no end to infinite possibilities, and Everything is possible. We can never have a total experience of Infinity. Even God cannot have this experience because reaching and fully exploring infinity is impossible. Infinity is not actuality but absolute potential. Infinity, understood in this way, is the ultimate Source of the World's freshness, uniqueness, and meaning.

UNIVERSAL MIND: THE ENGINE OF EVERYTHING

If the gravitational field is space, according to Einstein, then this field is the generator of Everything. What makes the gravitational field or space function is motion or the original, primordial "vibration" (Max Planck's term). Without the original vibration, there would be no motion and no gravitational field. Without motion, there would be no gravitation, no gravitational field, no space, and no curvature of space.

The main question is what powers matter and motion. The answer is that the engine of Everything is the Universal Mind. Matter is only a "program" of the Universal Mind, which we perceive as matter based on our sensory apparatus.

Time is not possible without space. Everything beyond the "physical" world is either endless or eternal. Absolute void is endless and, therefore, nonexistent. Absolute void is space without space. In absolute space, every point is the same point. Absolute space is pointless and spaceless. A pointless point is out of space or place.

There is no time in eternity since there is only the present or now. Absolute time is, therefore, nonexistent. Absolute time is time without time. In absolute time, every moment is the same moment. Absolute time is timeless. Timeless time is out of time.

Only in the created world is life and evolution possible in a real sense. Space creates distances and relations, and time is another type of measure. In this sense, the world would not be possible without space and time. The World beyond its "physical" realm is not measurable. Only things with form can exist, coexist, relate to one another, and have a duration.

According to Einstein, gravity is not a force but the shape of space curvature in the spacetime continuum. If gravity is not a force, and it is not, but the effect of the curvature of space, then the curvature of space is the effect of motion. This spacetime continuum and, consequently, the curvature of space would not be possible without spaceless space and timeless time. Spaceless space is infinity, and timeless time is eternity. Spaceless and timeless are the basis for space and time. Universal Mind is the basis for creation. This creation is the creation of the World ("matter"), space with curvatures, and time that can be "measured." Without an absolute void, there would be no space, and there would be no time. Without space, time, in its real sense, is not possible.

At the point of creation, two poles of the Absolute, Universal Mind (Something) and Absolute Void (Nothing) become creators of the world. The Universal Mind is the active "pole," and the Absolute Void (Nothingness) is the passive "pole." At the "moment" of creation, the Universal Mind envelops nothingness by the size of its program, but there is always nothingness "beyond" this nothingness unaffected by the creation. There can be many of these creations or universes. All of them together form the Universe. The Universe we live in is only a speck of the real Universe we cannot identify based on our current technology. It is a big question whether we will ever be able to glimpse the Absolute except through abstract thought and imagination, which are, in my opinion, legitimate methods not only of philosophical investigation but also scientific ones.

Without the original impetus from the Universal Mind, matter would be lifeless, immobile stuff, even if it somehow existed independently, and space would only be emptiness or absolute void, unaffected or minimally affected by the created world or universe. What powers the Universe is the main force, the source of all forces, and Everything. All the rest is an illusion, and our interpretation is a consequence of our senses and thinking. Our world is a program of an Absolute, Universal Mind working with the Absolute Void (its only means) to create the World. The main unifying force enabling all this is the Absolute itself.

The Universal Mind is the Perpetual Motion Engine and, as such, the Source of existence. The words of David Bohm are symptomatic:

> "In considering the relationship between the finite and the infinite, we are led to observe that the whole field of the finite is inherently limited in that it has no independent existence. It has the appearance of independent existence, but that appearance is merely the result of an abstraction of our thought. We can see this dependent nature of the finite from the fact that every finite thing is transient.
>
> Our ordinary view holds that the field of the finite is all that there is. But if the finite has no independent existence, it cannot be all that is. We are thus led to propose that the true ground of all being is the infinite, the unlimited, and that the infinite includes and contains the finite. In this view, with its transient nature, the finite can only be understood as held suspended, as it were, beyond time and space, within the infinite.
>
> The field of the finite is all that we can see, hear, touch, remember, and describe. This field is that which is manifest or tangible. The essential quality of the infinite, by contrast, is its subtlety, its intangibility. This quality is conveyed in the word 'spirit' the root meaning of which is wind, or breath. This suggests an invisible but pervasive energy, to which the manifest world of the finite responds. This energy, or spirit, infuses all living beings, and without it, any organism must

fall apart into its constituent elements. That which is truly alive in the living being is this energy or spirit, and this is never born and never dies."[4]

[4] David Bohm wrote these words in 1987 to be read at the memorial service of an old classmate from Penn State, Malcolm Lee Sagenkahnand.

CHAPTER 2

MIND AND MATTER AND MIND-BODY PROBLEM

One of the main philosophical problems has always been the dualism between mind and matter. Although logical and rather obvious, this dualism becomes less logical and evident if we break the limitations of our senses and habitual thinking. This paradigm has established itself as almost the absolute truth, but is it true that there is a definite difference between mind and matter, or is matter only a property of mind, a creation of mind?

I believe that matter is not only a creation of mind or property of mind but a manifestation of mind in a tangible form as a means of creation, recreation, and preservation of life, which is the fundamental purpose of an otherwise abstract and almost sterile mind.

Since we cannot think from a different frame of mind or perception except the one given to us by nature and established paradigms, we can hardly rationally comprehend where this dualism's secret lies or that there is no absolute dualism.

Matter, as something obvious and tangible, is all around us, but as we already know—what we perceive is only a presentation of the world to our senses through which we make the picture of the world. Our senses allow us to touch, smell, and feel love, pleasure, and pain. Everything looks material in a literal sense, yet what lives is only an idea translated through the senses. All we feel, touch, smell, and see exists because we not only touch it, smell it, and see it but are aware of it and have an idea about it. Beyond all we see or imagine, there is an overpowering mind. Without this, mind-matter is not only dead but is also not possible.

Matter is a manifestation of a Universal Mind and not something inherently different from the mind. Accepting our limitations allows us to see and understand what matters. In ancient times, the difference between body and mind was less significant to ancient than modern man. In this way, we become more perceptible and tuned to higher frequencies that lead us to ever creating and self-recreating Mind through matter as its program of life.

If we could become accustomed to this kind of reasoning, we could recognize that what determines our usual way of thinking, not necessarily perception, which we often condition by thoughts, is the paradigm accepted as "absolute truth." However, it may not be the truth. It is impossible to imagine anything as absolutely dead except something nonexistent, which is, on the other hand, not dead but only nonexistent.

There is no actual death in the Universe, only a transition from one state to another. The fact that matter, for the most part, does not have an awareness of itself does not change this fact. Matter itself, in all its forms, is alive everywhere. This life is possible only through something which channels it and feeds it. That something is the Absolute Mind or what, often misused and misinterpreted, the word God means in a deeper and broader sense.

We must use certain words; it is the main way to express our thoughts. Even with simple words, like mind and God, the matter becomes more complicated when we use them as terms with specific meanings outside their usual meaning. In this sense, God, as interpreted in most religious books, except in Buddhism, is not only personalized but becomes personal—he "listens" to and "cares" about us in a literal sense. On the other hand, the Absolute Mind is a different idea of the Being. However, what exists exists, regardless of our interpretation, and is not affected by our interpretations, except as a development of ourselves, which is, at the same time, the life of that of which we are part—of the Absolute Mind.

MODES AND QUALITIES OF REALITY
PRIMARY AND SECONDARY QUALITIES

Questions and debates related to the distinction between primary and secondary qualities, starting with Galileo, Descartes, Newton, Boyle, and culminating with Locke, Berkeley, and Hume, although we can go back to Democritus and his conventions, arise not only from these qualities *per se* but also from the lack of clear and precise definitions of these terms, including the terms "sensibles" ("sensible qualities") and "proper and common sensibles." For the philosophers of old, since Aristotle, proper sensibles were the same as secondary qualities for the philosophers since Locke. Common sensibles would be primary qualities based on Locke's classification. The main distinction shall be sought between the essence of the Being as a singularity, in its ultimate mode, and its manifestation, appearance, in and through plurality. We can further postulate that there is a distinction between the essence of singularity and its appearance or manifestation in (through) plurality.

The next question is whether Plurality saves the essence of singularity. Although singularity is saved even in plurality, this essence hides beyond appearance, and the senses cannot experience it. The senses experience only the appearance of plurality, not its essence as a singularity.

If we use the same terminology the empiricists used, we would conclude that the only primary quality is the essence of singularity, or the essence of the Being as it is and not as it is presented to us or perceived by us, and that secondary qualities are those as imagined or presented and sensed by our senses. Understanding primary and secondary qualities in this way leads us to the only distinction between the Being, which is in its ultimate mode, immaterial, and its manifestation of the World as we see it. Even though the Being, in its ultimate mode, shall be understood only as a singularity, its essence, being saved in plurality as its primary nature, is not annihilated through the creation or recreation of plurality and reality as we see it.

Therefore, Locke's distinction between primary qualities of the thing, which he described as solidity, extension, figure, motion or rest, and number, are not primary qualities of the essence of the Ultimate Being as it is but, at best, can only be, conditionally speaking, primary qualities of the manifestation of the Being in things, in plurality. As such, the distinction between the primary and secondary qualities is not between the essence and appearance (reality and appearance) but between the different modes (levels, properties) of appearances.

In its ultimate mode, the essence of the Being is beyond human perception and cognition. Whatever is presented to the senses or can be experienced or inferred from experience and knowledge is not the underlying essence but the appearance. So, when we talk about primary and secondary qualities, we are instead talking about the degrees of qualities and properties of the world as we "see" and sense it, not as it is.

Analyzing secondary qualities, as described by Locke, the color, temperature, smell, taste, and sound of the thing, we can see that primary and secondary qualities differ in a degree rather than in quality, which is to say that they are various degrees or modes of qualities (in a way, various degrees of the same ultimate quality) expressed and manifested by the Ultimate Quality (Universal Mind) of reality in the form of the world. Even if we accept this distinction, these qualities can only be the primary and secondary qualities of appearance and not the essence. They are various qualities (categories) within the same realm of the "material" world.

The real underlying, primary quality of existence—the world, the Universe—is immaterial and cannot possess any property we usually ascribe to the primary or secondary qualities. This higher reality, the ultimate quality of reality, is beyond our reach and is described by Plato as noumenon. According to Plato, noumena are the ultimate source of human knowledge. Kant thought that the noumenon is beyond human understanding and cognition. While the thing in itself (*das Ding an sich*) is mainly interchangeable with *noumenon*, they are not necessarily the same in Kant's philosophy.

Let's take the freedom to postulate the "primary quality" of the ultimate essence. The only distinction can be between this "primary quality" (essence) and the quality of its manifestation (appearance) of the world (material universe) as a so-called "secondary quality."

The heat source is the feeling of heat, not the heat itself, because there is no heat except an external impulse or source causing the feeling of heat. Likewise, external sources affect any other sense in how it reacts. Similarly, the source of cold is the feeling of cold, not the cold itself, because there is no cold. Anything we see does not exist as such. Everything we see is only the states of "mind," innumerable states of a living mind. Everything we see is the living mind telling us that what we see is exactly what we see. Since there is no color, shape, heat, or cold, we all experience this through the power of the living mind that gives color, heat, shape, and characteristics (both primary and secondary qualities) to the world through us. However, what enables us is the Ultimate Primary Quality of all, pervading and empowering all existence, to a larger or lesser extent, including human beings.

Since what we see is the product of a mind, an "immaterial substance" and not an actual state of the world and matter, then what we see is an "illusion." Without this "illusion," no sight, thought, or life would exist. The mind, or spirit, is the creator of the world and, as such, the creator of matter. Matter, as we experience it, is nonexistent. Matter exists as a state of mind or spirit in its numerous states, able and capable of creating, recreating, and preserving life—the most beautiful "illusion," more real than reality.

Matter is "immaterial" and a program or convention of the Supreme Mind. Physical laws, or laws of nature, unknowingly try to decipher this program or convention.

To understand matter means to understand life.
To understand life means to understand its purpose.
To understand purpose means to understand the meaning.
To understand meaning means to understand the world.

If and when we understand the world, then we understand the mind. When we understand the mind, we understand matter, and there is no dualism between mind and matter. We understand that all we feel and see is inside us, a living mind in action.

The matter we see is not the matter that exists. The whole Universe is a vast transmitter of the ultimate message of life. Every being that participates in the game of life is a messenger, regardless of how minuscule it is. The whole Universe is a mental system. This system powers all that exists. Perception is an idea. Matter is an idea. There is nothing beyond thought, mind, spirit, or information.

Qualities of matter shall not obstruct us from "seeing" the objective (actual) reality. Weight, height, and all properties of matter—mass, motion, radioactive energy, electromagnetic energy—are properties or emanations of mind, not matter. We give them these qualities; they do not exist *de facto* or *per se*. All these states of mind, programs of mind, exist, but what we see and experience is an "illusion" of a living mind. Without this "illusion," there would be no life, and the purpose would be lost.

All elements from the system of elements are different states of a living mind. There is no hydrogen, helium, gold, or zinc ... What we see is a convention of our minds programmed to see or to have an idea about what they see. It does not mean that all this does not exist; it only means our perception is so distorted that we do not understand its real nature.

Someone may say, if it all does not exist, except in some other way as a mind or spirit, mental system, then how can we be killed by some material object and feel or suffer all other consequences of the material world, including pleasure and pain? All this and all the consequences must be there, which doesn't change the fundamental nature of it one bit because the program of a living mind causes, produces, and preserves all the properties of a "material" world, and all consequences, expressions, and feelings among others, that come from relationships within a world as a living mind "materialized" in action. Outside of this world, these consequences may cease to exist because there is no residing mind in action to create and preserve an "illusion."

Matter is not "real" but a convention of our senses, minds, and feelings. (Plato intuitively felt this, although on an abstract, metaphysical level, which explains his fifth element in *Timaeus* or Aristotle's idea of aether.) All laws of nature are conventions of a living mind in action. Since there is a Being, the Being has to have its nature. Only Nonbeing has no characteristics. Still, this does not mean that matter exists as we see and perceive it. The main feature and quality of the Being is non-material, in our sense of the word material, and is first and foremost mental; it is information, knowledge, thought, Mind (we use these terms in the broadest possible sense). Although our mind cannot function without matter as we see it, it is still the source of what it perceives, not its result. Dualism is yet another beautiful illusion of the senses.

DEGREES OF QUALITIES AND REALITY

Immaterial essence is the primary quality; its "material" appearance is the secondary quality (primary quality to Locke), while the sensations are tertiary. The primary quality is the absolute quality of the Being (present yet immaterial and invisible). The secondary quality is the world as unaware information (unaware "awareness")—perception and awareness cause the tertiary quality. Secondary and tertiary qualities are the result and degrees of the living mind (immaterial essence) in action.

Matter is nonexistent except as something we see and experience as matter by our senses. Matter is only a manifestation of a Universal Mind. The primary quality of matter and all existence is not accessible directly or indirectly except by pure thought (intuition). All qualities, described by some major philosophers as primary and secondary qualities, I understand only as different modes of existence and relationships between the sensed or perceptible and different modes of perceptibility or perceptions (and senses) of the beings possessing these abilities. These qualities represent different levels (manifestations) of matter (existence and life) and do not disclose the actual nature either of matter or true primary quality, the underlying reality of all, which to me is only something I chose to call a Universal Mind. This Universal Mind, in some ways, may correspond to Plato's (or Kant's) noumenon, to Aristotle's aether in a way, to Parmenides' Mind (Oneness), to Anaxagoras' nous (except that there were no things mixed up together before the mind or nous arranged them; instead the "things" are only potential or the transformed mind, manifestation of a mind), and to Plotinus' ideas of a mind of this kind. Heidegger states that philosophy, since Plato, has neglected the idea and the Being itself and forgot to ask what it is.

There is a world beyond the world, the primordial Being, which I call the Universal Mind. This world is independent of its creation, yet its creation depends on it. Still, in its metaphysical "form," the Being loses its purpose without creating or recreating. The Being must create to keep itself alive. Without creating, it is dead. Creating is not only the source of the Universe but also the source of the continuous life of the Being through myriad forms. The life of the Primordial Being is dependent on its creation and *vice versa*. The life of one depends on the life of the other, although they always, in a way, stay the same. The Creator is its creation, and its Creation is the Creator. The Universe, or the World, depends on the Primordial Being, the Universal Mind. Yet, it is conditionally independent in creating the always new Self, a new World (ultimately the source of free will).

The Creator, the Primordial Being (Universal Mind), is the Ultimate Primary Quality of Reality. Creation (energy, matter) is the secondary quality of reality. The effects of the secondary qualities on the mind are tertiary qualities on my scale of qualities. In Locke's classification, our secondary quality would be primary, and our tertiary quality would be secondary. This distinction is essential for understanding the nature of reality.

There is no energy or matter as a physical force of nature *per se*. We experience the "embodiment" (our Reality) of the Universal Form as a result of the conditioning of the Creation (particular beings) to perceive it as matter and energy based on the Ultimate Program of the Universal Mind *via* the tertiary quality of reality. Being conditioned to perceive something as matter or energy, as hot or cold, as blue or red, we will perceive it as such based on our conditioning, not necessarily based on the true nature of energy and matter. Reality, energy, and matter, as such, do not exist. Reality exists as presented to our senses. What we experience is not the result of reality but our conditioning to perceive reality as such. How we see and perceive reality comes from the third quality of reality, without which everything would collapse. The whole of Reality is the Web of information. Thanks to the tertiary quality of reality, we are partially able not only to enjoy this marvelous Web of information and beauty but also to decipher it.

PROGRAMMING OF MATTER

Since almost everything we see (or do not see) is a matter of "convention," we can say that all we see is the result of our perception. The picture of "reality" and the world depends on our perception. Although we all see the same things (with slight differences measured in nuances), the same colors, and the same shapes, this does not change the fact that it all results from our perception. On the other hand, we are not the creators of our perceptions but the beneficiaries. The only difference we make is in the endless nuances and possibilities despite the limits of our perceptive powers.

Senses make perceptions possible with feelings, emotions, and thoughts in humans. From an earthly point of view, senses empower the higher levels of life, conditionally speaking. However, what empowers the world is the possibility of and for relationships. If the absolute world is ultimate and one, relations are not possible. Any division of oneness is, in a way, the opposite of the absolute. The Absolute must stay intact to be absolute, but the Absolute is "dead" at the point of its absoluteness, which is Zero. Absolute nothingness is dead at the Zero point of its absoluteness. The absolute Being (God) is "dead" at the point of its absoluteness, which is Zero. The whole Triad of the Absolute, with its components of God and Nothingness, is Zero if it is only the Absolute without established relations within itself.

The only exit for the Absolute is through the Being in a dynamic relationship with the Nonbeing or Nothing.

Omniscience, or absolute knowledge, annihilates knowledge.
Omnipresence annihilates presence.
Omnipotence annihilates potency.
Only the relative is the actual creator.
The absoluteness of the Absolute is its death penalty.
Absolute power is the annihilation of power.
The absolute power of the Absolute is its "weakness."
The absolute accomplishment is the ultimate failure.
Absolute life is the annihilation of life.
Life without death is impossible.
Life without death is immoral.

Absolute knowledge, or omniscience, of itself and without the world is the annihilation of knowledge and, therefore, the source of its demise. The power, dispersed in plurality, is the source not only of life but also of happiness. The Source is the ultimate power.

The only exit for God is life, but God cannot go anywhere; anywhere it goes, it will stay in the realms of the Absolute. Nothingness (we may say) is the opposite of God, its antipode, "nemesis."

Nothingness is emptiness, darkness.
God without the exit is a dead God.
God, without darkness, is "darker" than darkness.
God, without motion, is asleep.

Without emptiness or nothingness, an omnipotent God is impotent. The potency of God lies in its ability to live through the relationship with emptiness (nothingness, darkness).

Without the nothing, God is nothing.

Without emptiness, God is empty.
Without division, God is indivisibly dead.
The divisibility of God is the source of life.
However, how can it be divided if there is absolute oneness?
Absolute is the biggest enemy of the absolute itself.
To live, the Absolute must transform into relativity.

The Absolute preserves its wholeness within the infinite limitations of relativity, which feed the Absolute's absolute power and endless possibilities.

The limit is the source of infinity.
Without limits, there is no infinity.
Limits establish forms and, through forms, relationships.

Everything in a relationship with something else is relative to that else and all the rest, yet its relativity has absolute value. Nothing that exists can be without its intrinsic value. The intrinsic value of everything that exists is absolute.

Every part of the Absolute is connected and interconnected through the intrinsic value of everything that exists. The absolute value of the finite is absolute in its finitude and infinitude. Its finiteness makes connection and relationships (life) possible in the absolute sense. The infiniteness of the finite is absolute because it lies in the ability of the finite to resemble the infinite in its potential for variations. Every finite value is potentially infinite.

The Absolute, without life, is dead not only because of its power, omnipotence, and omniscience but also because of its absolute finiteness. The Absolute is the most finite thing. The totality of power and knowledge makes the absolute obsolete. Its total knowledge is its ultimate limit. Its total power excludes purpose and meaning. There is no purpose and meaning in absolute knowledge and power. The purpose is already there in the Absolute, that which should be longed for.

There is nothing beyond the Absolute. There can be no purpose, meaning, or hope if there is nothing beyond it. Purpose, meaning, and hope can only exist in relativity. Without relativity, these terms or states have no meaning. Relationships create purpose, meaning, and hope. Whenever or wherever there is a relationship, it means there is something beyond, something more significant to hope for, something to pursue and build meaning and hope through that pursuit.

Absolute is its own purpose, meaning, and hope. Since it cannot pursue itself or enjoy itself to establish meaning and hope, it must establish, revitalize, and empower the first and the most important relationship—the relationship between the Being and the Nonbeing—the relationship between God and Emptiness, light and darkness.

Everything in nature coexists with something else. Our perceptions come from relations. We must see and perceive what we experience in specific ways; slight differences don't count. What we see is not a matter of choice for the most part—to see the world, to feel heat, cold, and fear. All that exists has its frequency and structure within a larger structure. Since there is a connection between everything, the existence of one depends on the effect it makes on another in a very peculiar way. This effect is more of a result when the impacted one is the source of life and existence for the One that initially caused and produced it all.

MATERIALIZATION OF THE UNIVERSE

The term known in philosophy as "primary qualities" of physical objects (solidity, figure, extension, motion, and rest) is, in Reality, secondary, and the term known as "secondary qualities" (color, odor, taste …) is tertiary. In my system of thought, the Universal Source (Mind, Spirit) is the Primary Quality of the Ultimate Reality, which programs the subrealities of the secondary and tertiary qualities (previously defined as primary and secondary). The point of transformation or passage from the Ultimate Primary Quality of Reality (the Being) into the secondary quality is the point or moment of materialization. That is the moment when the Universal Source (Ultimate Primary Quality) "creates" the secondary quality of the "physical" world that previously was defined as the primary quality. That is the moment of creation of countless interconnected webs. The "new" world is plurality, as opposed, on the surface, to the initial Oneness (singularity) of the Ultimate Primary Reality. Nevertheless, the underlying Oneness of Reality is never lost. In the world of plurality, where everything affects everything else, the Unified Field of Reality is created for the tertiary quality (previously known as secondary) to function upon impulses and signals of the secondary qualities.

In such constellations, not only do the secondary qualities affect the tertiary in the form of "impulses," but they trigger reactions in tertiary qualities without which the secondary qualities would lose strength and be, in some ways, almost nonexistent. Existence, if it is not aware of itself or not recognized, can hardly be characterized as existence in a more profound sense. Without the primary quality, there would be nothing. Without secondary (originally primary qualities), there would be no tertiary qualities (originally secondary), but without tertiary qualities, secondary qualities would, to a large extent, lose meaning. Interdependence among these qualities is such that the disappearance of one almost automatically means the disappearance of the other.

Through the process of "materialization," the Universal Source "inseminates" and "fertilizes" emptiness (original void) in the love-making cosmic hug between the Universal Source and Emptiness. Reality, even the one we can experience to some extent, is mind-boggling. There could be up to 2 trillion galaxies in our Universe only and up to 200 sextillion (200 billion trillion) stars.

In this magnificent and complex web of information, everything affects everything else. It is "programmed" in ways that the objects of "physical" reality have some properties and features that affect tertiary qualities in a way that tertiary qualities recognize them with all their features. Depending on the level of tertiary qualities and dispersion, all the feelings, measures, weight of objects, pain, and anything we can imagine will be recognized and felt by tertiary qualities as a direct impact, effect, and consequence of secondary qualities in a literal sense. For instance, people would feel and be able to measure the weight of a physical object and all the rest and be able to feel pain, among other things, as if the whole energy and matter and all their properties and effects were real and not the "programmed" product of the immaterial Universal Source (Mind). Even after the "creation" of the world as we see it, it is still immaterial. Yet, we perceive its qualities and properties based on the "program," which dictates our reaction to secondary (originally primary) qualities of the world.

Even if imagined matter existed independently but did not function as we described, there would only be dead matter. Matter, "artificially" made by the Universal Source, is still immaterial but is more real than if matter existed on its own, irrespective of the Universal Source or its impact. All we experience as energy and matter is immaterial information from the Ultimate Source of all Reality, without which there would be only absolute Nothingness.

NOUMENA AND PHENOMENA

According to Plato, the Ultimate Primary Quality (Noumenon, or the thing in itself, the Being, or God) is accessible by pure thought or intuition. Since it is not in the "material" mode of secondary quality (formerly primary), according to Kant, it cannot be accessed and experienced by the senses.

Kant's idea about experience implies that it is the only measure of appearances or "reality" as presented to us and not as it necessarily is. Reality, as we experience it, is an illusion. The world we experience is the world of appearances or the sensible world, and the world beyond our reach, beyond the possibility of experience, is the intelligible world.

Kant insists on experience in terms of cognition and understanding, which further implies that pure thought accessing the noumenal world is, in a way, impossible since there is no experience to check it. We state not only that the whole world is an "illusion" but that the Noumenon, as the ultimate source, is the creator of the phenomenal world operating as a program of the Noumenal domain of the same world.

The phenomenal world is only a different domain of the noumenal world. It is the "intention" of the noumenal to become, on some level, phenomenal. Although from the perspective of the phenomenal, noumenal seems to be metaphysical and transcendent, from the perspective of the noumenal, phenomenal is immanent. Regardless of not having direct immaterial access to the noumenal, through our experience of the phenomenal, we experience the noumenal at the same time.

In my system of thought, noumenal is the immaterial oneness or singularity, and the phenomenal is "material" plurality. Kant thought that the merging of phenomena and noumena transforms everything into appearances and that this would be the artificial way or "illegitimate" way to experience noumena. Since the created world is an "illusion" (conditionally speaking), everything stays noumena. Still, on the superficial level of the made reality, we experience the hierarchies and degrees of the qualities of the new reality.

The new "reality" requires different experiences and imposes other instruments of experience. The World beyond "new reality" does not require sensory experience. Universal Mind (we can equate it with Noumenon) is still the underlying reality, hypostasis of all, and our experience of "reality," or phenomenal, is, at the same time, the experience of the noumenal, but on the minuscule level. This ability of the noumenal to transform into phenomenal is the secret of existence, life, purpose, and meaning. We may say that, without appearances or phenomena, noumenal loses meaning.

The things in themselves (or the thing in itself) do not exist in the noumenal realm; they are only the potential of noumenal. The Noumenal exercises its potential in the phenomenal realm. There is no way to experience a thing in itself because appearances are not direct reflections or copies of ideal forms/ideas nor the appearances of things in themselves, but the expressions of the noumenal through phenomenal. Appearances do not have the corresponding reality of the things in themselves in the noumenal but are the products of the absolute potential of the noumenal.

This absolute potential possesses potential for unlimited variations. The possibility for variations is infinite. The infinite potential and possibility in its total capacity are unknown even to the Noumenon itself, and that is probably the main secret and beauty of existence, of the noumenal and of phenomenal, the main secret of the Being capable of becoming and capable of just being the Being in all its glory and potential.

When we use the word illusion, we mean something is not real in a broader sense. Yet, how do we define reality and illusion? Does reality depend on our senses, understanding, and definitions or on what it is objectively? Why would our senses, "definitions," and "understanding," or lack of it, not be reality irrespective of our idea of reality? What constitutes reality? Who decides what reality is? Would it not be logical that whatever exists is real? Even if something does not exist, we can imagine the reality of nonexistence. If everything that exists is real, we can only talk about the degrees or levels of reality without denying reality to something we do not understand. Our lack of understanding shall not be an obstacle to reality but a motivator to try harder and get closer to the most "real" of what is possible.

Only the world of Oneness and Unity (Singularity) is a universal reality without hierarchies. Still, the World of plurality and dispersed Oneness is the world of hierarchies, orders, degrees of life, understanding, and evolution. In a World of total Oneness, there is no evolution. Evolution and real life are only possible when they become less "real," when something leads to something else, when something seeks something else, when something is bigger than something else, and when we crave and desire. The purpose of reality is life. The real way of life is motion. Without motion, there is no life. What secures the motion, whether matter is real or a product of a Universal Mind, is less important than the ultimate purpose of it all, and the ultimate purpose is to preserve life and meaning against nothingness.

Indeed, we do not experience noumena directly. Still, since the phenomena are the program of the noumenon or the transformed noumenon itself, we may be able to think and understand (to some extent) the thing in itself, Noumenon or the Being (Universal Mind). Although transcendent, Noumenon is immanent at the same time. Phenomena are the emanations of the Noumenon.

Noumenon, or the "thing" in itself, is the reality without reality.
Ultimate reality is the annihilation of reality.
A Sourceless Source is sterile without creating.
Ultimate Source is the beginning and end at the same time.
Without creating or transforming, the Ultimate Source is the dead world.

The purpose and meaning are found in constant transformations and creation of multitudes, allowing the ultimate oneness, singularity, to breathe and live through dispersion into the nothingness that the Source swallows.

Nothingness is the "air" of the Ultimate Being and the air of countless beings dispersed in the new entity we call the Universe.

Our senses, cognition, and understanding are the result of conditioning. We are not the creators of our senses or our cognition and understanding in the deepest and fullest sense. Without our conditioning, there would be nothing. Senses, cognition, and understanding among human beings may differ only in degree, based on education or intellectual capacity, but not in mystical or mysterious ways.

We can follow Plato and Kant and agree that the world of phenomena is an illusion and that noumenon is reality. Still, we must add that reality is lost or undermined without this illusion. In this way of reasoning, we conclude that although reality is the creator of an illusion in the form of an "artificial" reality or the world, this "illusion" is also the creator of the reality of the Being itself or the thing in itself. Both reality and illusion are equally important. Without the one, the other loses its meaning and purpose.

Then, we may ask the question: What is reality? What is the real or objective reality? Finally, we may be surprised by the ultimate answer of reality: that the thing or reality is the illusion itself because the Ultimate Source, the Ultimate Reality, at its supreme point, is equal to Nothingness. That would mean that the Ultimate

Reality is Nothingness. The Ultimate Source is the Ultimate Potential. Whether the actualization of this potential is reality or illusion is irrelevant. What is important is the existence and realization of the potential.

In actuality, reality is an illusion. If it were not an illusion, to some degree, it would be the "God" itself, the realization, Oneneness without the beginning or end. That would end everything because everything would transform into its primordial state of Nothingness. Ultimate reality, or Nothingness, is therefore without purpose. The purpose is created by altering the Ultimate Reality (noumenon) into the world of plurality, so the plurality itself is an "illusion" that secures purpose. Without this illusion, there is no reality in an absolute sense. The road to reality is an illusion. Thanks to this illusion, there is existence in a broader sense. Limitations are the source of movement. Without separations and limits, there would be no movement but only frozen Oneness.

IDEA AND WILL

For Schopenhauer, the world is an idea. Although in a way distinct from the will, this idea implies will and therefore equates it. If the world itself is an idea and if nothing exists beyond this world and this idea, then there is no place in Schopenhauer's philosophy either for noumenon or metaphysics. If everything is the world and the world itself is an idea and the will, then the whole world is a phenomenon: subject and object, cause and effect, purpose and meaning. Although there is a distinction between the idea and matter, this distinction is only on the surface, since even if the world is an idea or an appearance of a hidden idea, this ultimate idea is not beyond the world but is the world itself, which annihilates the substantial distinction between mind and body (matter and idea).

If the will materializes as an idea, this distinction becomes less distinct. Almost the same scenarios, as in metaphysics, can be applied here within the realm of the physical world. What serves the role of the noumenon in Plato's sense (even Kantian) is replaced here not by a metaphysical (transcendental) idea but by an always-present "idea," carried by will and manifested through the world (matter).

If we equate Schopenhauer's idea to noumenon, this idea can become the mind's essence in our sense. On the other hand, if we equate Schopenhauer's will to the world, then we can equate it to the secondary quality of the primary essence (Locke's primary quality). This almost invisible dualism in Schopenhauer's thought can be easier to understand if we treat his idea as essence and his will (manifestation through the world) as existence. We can further conclude that without the essence, there is no existence but also that without the existence, the essence "disappears" (is on hold as potential).

Therefore, the will must be the fuel of the essence. Without the will, there is no existence; although it contains omnipotence, an idea still dies. In this sense, the will for existence and life is equally important as an idea because, without the will, the essence dies. The only way for an idea to survive and live is through existence. (Without the will, there is no existence; without existence, there is no world; without the world, there is no idea; without an idea, there is no will, which implies that idea and will are essentially the same.)

We can also "equate" Schopenhauer's will to Empedocles' love, except that Schopenhauer's will is more ominous. Ontologically or epistemologically, as ultimate principles, they do not differ in substance, only in form.

The difference between love and will, although only formal, is fatal regarding the worldview or outlook of the philosopher in question. Everything must be joyous for the one coming from the principle of love, but for the one coming from the principle of will, as a blind force and strife, the world is less rosy and cheerful. This pessimism comes less from two inherently different models that would imply two different sets of rules (happiness, among other things) but more from the inability of the world to get rid of its will, which is to say, to commit suicide, to get rid of itself. In the case of love, this proposition is impossible because that is a joyous and lovely world. Still, in the case of blind will, the main characteristic of the world is existence at any cost, regardless of hardship, misery, and pain.

Despite his originality, it would be hard to imagine Schopenhauer's will without the Kantian will. Schopenhauer "converted Kant's moral will" (moral regulator or reason) into a metaphysical will on a universal level.

(Also, it is possible to convert Kant's reason to an idea on a universal level to clarify the will's effect on the reason and its importance and meaning.)

Schopenhauer's will can be "equated" to the categorical imperative of the idea (reason), which is the will. Without the will, the idea is dead. If the idea is the essence of the Being, then the will is its manifestation as existence. Only through the categorical will of the essence (ultimate universal idea) is existence possible.

All suffering is the price of life. All suffering must be in balance with all beauty. The pessimistic view is more an expression of the individual state of mind (outlook) than the external state of affairs ("objective reality"). According to Kant, the will is the universal legislator. Similarly, we can equate pure reason to the Supreme Being (essence) and practical reason to a plurality or multiplicity (the world, existence).

(Kant thought that freedom of the will is only an idea and that, although it can function as a possible categorical imperative, it is still only a hypothesis. For Kant, it is impossible to explain how pure reason can be practical in itself, and, according to him, this is "beyond the power of human reason.")

Still, suppose this idea or hypothesis is analyzed or applied to our concept of the Absolute. In that case, this seems a rather obvious consequence of such a system.

For Schopenhauer, there is only one underlying reality; for Kant, there are things in themselves as a plurality. The difference is singularity against plurality (diversity). But this difference may be only on the surface, for it is hard to imagine that Kant thought of noumenon (if equated to a thing in itself) as of plurality, but rather that things in themselves are not differentiated in the noumenon as they are in the world of phenomena for these phenomena are only particular, phenomenal manifestations of the One—Noumenon (although this may not be the case with Plato).

Let's think deeper about Plato's idea of noumenon. We may conclude that, although on a superficial level, noumenon may contain plurality, when we look deeper, we may conclude that Plato's noumenon is singularity too. Regardless of the description and explanation in the *Republic*, Plato's noumenon is or may be the undifferentiated One. The idea that the world we see and the things in it are only the shadows of an underlying reality or noumena does not necessarily mean that all these things have their literal equivalents in the noumenon. In the end, there seems to be less difference between Plato's forms (ideas) and Kant's things in themselves than it looks like on the surface. Still, noumenon, although being a singularity, being the One and universal underlying reality, contains plurality as a potential.

TESTABLE AND NON-TESTABLE NATURE OF THE UNIVERSE

Although instruments can be almost 100% accurate, instruments are limited when we deal with the nature of the Absolute and the darkest areas of the Universe. Exact sciences, especially physics, rely on experiments, but there is a limit in the human mind and nature, which is relative too. To comprehend the Absolute through the relative, only from the point of the relative, is impossible. To test what is beyond the testable nature of nature is impossible. Absolute cannot be subjected to experiments. Absolute is not measurable or definable. Absolute escapes any definition based on strictly material existence or laws of material nature.

The only measure of the Absolute is the Absolute itself. Absolute is an immeasurable *mind.*[5] Since everything, regardless of being dispersed into relativity and plurality, is part of the absolute mind, then everything intrinsically possesses a part of its nature. Regardless of everything possessing a part of the absolute mind, not everything has a mind that is aware of itself. There is a hierarchy in the world of plurality. The dispersed reality of the Absolute is not based on the human understanding of matter, energy, thoughts, feelings and emotions, laws of nature, qualities, and levels of awareness.

---

[5] We use the word mind for lack of a better word, so this Mind is not only superior to a human mind but different, and we use it more because of its immaterial nature than because of its similarities to a human mind.

Philosophers and scientists throughout the ages have been concerned with these questions. Still, the question is not only about posing the question but also about posing the right question and understanding the meaning of words and language. The right questions and good reasoning often lead to the correct answers, but we are dealing with how and when we establish the right concepts. Have we ever? How far are the concepts, beyond our words and language, from the intrinsic nature of what we try to describe and comprehend?

The only measure of this is the truth. But what is the truth? Is the truth what we establish as truth or what it is, in its absolute meaning, independent of us? Does the truth of language, thought, and meaning wear the colors of our understanding? If the truth, as we see it, wears, to some extent, our colors, the truth will always be, at least a bit, distorted.

Regarding the old question of the mind-body relationship, the concepts related to this topic have changed from ancient times, and the difference between body and mind is more evident to us today than it was to ancient man. What changed? Was this development a result of growth in understanding or a simple paradigm shift? Is a paradigm shift always connected to growth and better understanding or not? All paradigm shifts must rely on concepts. These concepts are established based on ideas. All ideas could be better. There is a perennial fight among people about ideas, among other things. Since ideas are not easily measurable, they can be established and, based on them, the rule, regardless of their true merit. Ideas, irrespective of their intrinsic value and truth, can become the inherent values of society and may even become the "truth" itself.

To reach the truth, it must be at the absolute level. But does the absolute truth exist, and what does it mean? Do ideas represent truths? To what extent do ideas represent truths? These questions mostly relate to society and abstract or concrete questions concerning ethics, aesthetics, psychology, philosophy, and religion. Exact sciences are based on and governed by different standards and concepts of truth or ideas about the truth. Regardless of this dichotomy, it is only a dichotomy on the surface. Deep down, the absolute truth is at the equidistance from all these essential points, or all approaches, regardless of their origin (based on purely theoretical thought or conclusion resulting from an experiment), provided that all these approaches have equal merit based on the intrinsic value of any particular endeavor or approach.

CHAPTER 3

DIMENSIONS ARE CONVENTIONS

As we understand it, space is three-dimensional: length, width, and height. However, do these words or terms represent dimensions? If we think a little deeper, we conclude that they do not since the length, width, and height are the features and properties of space, not its dimensions.

What is, then, a dimension? Dimensions are measures of specific properties, features, or states of space. For example, a length of 20 feet, a width of 15 feet, and a height of 10 feet make up the space of a particular room or its dimensions. However, only the length, width, and height make up space, while dimensions represent its measure.

Length, width, and height are not dimensions *per se* but properties of space, with dimensions presented to our senses rather than as they are in their essence. If we say that some space is 10 feet long, we know one of its "dimensions." Dimensions are measures of space and not its properties. Length, width, and height are the properties and features of space, and their quantitative value represents dimensions. Dimensions are our measures.

Thus, we can measure space as something formed with its properties. Still, space itself is not dimensional because dimensions are its measure—a sensory measure, not a property. Space in itself is nothing, and nothing is non-dimensional. Somebody who weighs 200 lbs. is not weight but something with "weight." Somebody who is 6' tall is not a dimension or a feature, although height is a biological, physical feature that we measure as 6’ height based on our perception.

Dimension gains its value only when something is already measured. Before we measured it, we did not know its dimensions. In this sense, we cannot say that space has dimensions but that specific quantitative values can be measured. Here, too, we come in contact with the idea of the concept, where the concept goes astray from an idea or truth in proportion to its linguistic separation from the rule or its original idea of the very words and their meanings. The linguistic purpose of the word dimension, originally, was to represent the measure of some of the features, or all, of space and not to be the very feature or property of space or of that to which the measure, dimension, is applied, or of that which it sustains.

The question is, the measure of what do dimensions represent? What is space? If we measure the length, width, or height of anything, we measure what is presented to us, through senses, as shape. Every visible shape in nature and, most likely, invisible too, is, for the most part, emptiness or nothing. That which we measure does not exist in a higher reality but is emptiness. That which gives a quantitative value to space is emptiness, not matter. If we could expel emptiness from space, it would lose length, width, and height. We measure emptiness, not matter, and emptiness is not dimensional; there is nothing to measure; it is the same everywhere. Something must exist to be measured.

The Reality (Universe, objects, matter, space) we measure is the Reality created by the Primary Quality of the Ultimate Source. When we measure the secondary quality (formerly primary), we are convinced it is as material as it appears. Yet, it is the reality programmed to be perceived as such by tertiary qualities (formerly secondary). Regardless of this “wrong” opinion or “perception,” Reality is not less real than we thought it was. It appears to be as it is, regardless of our views and acquired knowledge about it. We can and should adjust our worldview accordingly, but that does not affect us adversely. The change of perception should not necessarily be a drastic or detrimental worldview change. Quite to the contrary, this change shall boost progress and better align us with the cosmic organism we are part of. There is no reason to doubt or question reality, but we have to get used to a new, changed idea or understanding of reality since reality is not what we thought it was. Matter is not what we thought it was. Space is not what we thought it was. Our Reality is the Ultimate Web of Information. Whenever we measure something, we measure the information, the program, not space or matter *per se*. The only volume of Reality comes from Void.

If we think in this way, we realize that a dimension is a measure of something wrongly used as the property or feature of something, but space itself is not, in its essence, what we think it is. We imagine space, conceptually and linguistically, as something solid and conditionally emptiness if this emptiness is within something tangible as matter. In this way, all we measure is the measure of a shape that we see in the way we see it, not how it is. We can also measure distances between the shapes. This is how we form our idea of space. Even if there was no absolute vacuum, what gives space or dimensions to anything we measure or see as space in the "material universe" is this void or nothingness. We experience and measure all the physical qualities of reality, but they are only "coordinates" or informational, immaterial skeletons of Reality appearing in the forms we experience as "physical." We measure this very void for, without it, space or our idea of space is impossible.

SPACE AND ZERO

What makes space possible? Just of itself, space is nothing. But without nothingness, there is no space. Space is made only from nothingness. Without Nothingness, the Absolute would not be complete, and there would be no potential for space in the world. Absolute would be impossible without nothingness. Absolute, without the void, is zero, and the void, without the Absolute, is zero. Zero can only be equal to itself, which means that even in the supposed separation of the Absolute from zero, they become the same and only one—Zero.

Only by the Nonbeing does the life of the Absolute become possible. Without emptiness, motion is impossible; without motion, no real existence is possible. Only that which moves on a journey, exposed to and capable of change, can exist and live. Without the Nonbeing, the world is the Absolute which sleeps. The Absolute, which sleeps, is the same as nothingness.

Awakening of the Absolute is impossible without nothingness or emptiness. Here, we see the natural force of duality or relativity in action—the relationship between the Being and the Nonbeing as the only possible relationship by which we can explain, or try to explain, the world and its awakening, 'birth,' motion, and life.

Energy is the world-born phenomenon, the world that puts itself in motion and flies into space by receiving space into itself. From this point of view or the point of view of the Theory of Relativity, matter is indeed condensed energy. But, from the Absolute, or the Theory of the Absolute, both energy and matter are the dissolved forms of the primordial world of the Absolute. Therefore, energy is the "dissolved" Absolute, and matter is the formation of "energy" into objects of the multitude of the Absolute, which transforms from oneness and singularity into plurality (although oneness is never lost).

What is the Absolute? Absolute can only be the absolute knowledge, thought, mind, immaterial "substance," the absolute "brain" (spirit) that contains an infinite programming ability and potential for absolute plurality through chance. Since human beings are limited, although much more advanced than other animals, they cannot understand inconceivable and invisible spaces by experience, experiment, or evidence, far exceeding their perceptive and sensory abilities. Even if they were at a higher level of evolution, humans would be unsure if they found the final truths and complete knowledge. However, if they free themselves from dogmas—religious, political, and others—human beings can count on a higher degree of understanding and closer touch with the world.

One factor that obstructs our understanding of the world is our insatiable desire to understand it at any cost. If we insist on understanding the world at any price, we will not understand it or will only partially understand it. Such an attitude is motivated more by self-promotion than a desire for fundamental understanding. If we freed ourselves from all possible chains, we would expand our views, and what we wanted to achieve at any cost could be achieved at a lower price.

These few digressions serve as the basis for the easier understanding and acceptance of the Absolute as mind—idea (spirit), knowledge, thought—not only as energy and matter in how we see and perceive them with our limited senses and thoughts but unconditionally.

If the Being, Universe, appears in the Nothing, or the Nothing appears in the Being, a point in space is created, and consequently, time. What makes the Nothing a real nothing is the lack of something, the lack of the Being. If the Being appears, nothing is not nothing anymore, for it becomes space.

If there is only one primordial Being "in" the Nothing, there is no relationship between this Being and anything else, and all this void is equal to zero or infinity. But, if there is an appearance of a "different" something, the Universe, the "material" Being "in" the Nothing, there is a relationship between the two entities, and anything in and between them is space. On the other hand, since these universes are pluralities, there is already established space inside them due to the plurality and distances they cause.

Although the Nothing is nothing, of and for itself, it is not nothing when interacting with the Being. This interaction activates the passive feature, void as a potential for space, of the Nothing through relationships, and it provides the medium, space, for its division and expansion into existence. In this way, the Nonbeing creates the Being partially but equally important. On the other hand, the Being transforms nothingness from nothing into real space.

Absolute time is eternity. (Eternity = Present)
AT = Et (AT = absolute time; Et = Eternity)
Absolute space is infinity.
$AS = \infty$
$AT = \infty$
$AT + AS = \infty$
$\infty = 0$
The Being, the Universe, is finite.

SPACE

Although primordial emptiness is nothing, there is no space without emptiness: emptiness has no dimension. Only the compressed idea, compressed knowledge of the Being, hides "dimensions" within itself, but in its compressed mode

> It is beyond dimensions,
> It is beyond space,
> It is beyond time,
> It is the whole of compressed space and time.

After the voyage, every form comes to the same place despite being dispersed in myriad directions in conditional "separation" and dissolution. When forms completely dissolve, the emptiness separating forms or shapes becomes a primordial desert of nothingness without the shapes. This nothingness is the same everywhere; every "place or point" becomes the same "place," a pointless point; the distance disappears. Since there is no time without space, the same applies to time. Real space is nothingness or an absolute vacuum; it is non-dimensional.

In a singularity, there is no space and time; time is a consequence of space. Space and time are only possible in plurality. The irruption of singularity into plurality is the cause of space and time as we perceive it.

As we see it, space is possible only when the primordial Being swallows emptiness. Space is born when the Being accepts emptiness, the Nonbeing, into itself. Thus, this condensed Being begins to dilute and spread out; emptiness starts making relations within it. Its life is impossible without emptiness: vacuum gives dimension to the Being, although without any dimension. Emptiness provides volume to the Being.

Atoms (electrons, neutrons, quarks, strings), all we can imagine, are only messages and sparks of condensed knowledge and reminiscence. Emptiness gives the power to "matter," enabling it to transform and grow. The very fire of hydrogen is knowledge. The Universe is the purest in the moment of space-birth; its primordial fire glows from the first "element." "Fire" cannot evolve into something different from its existing knowledge. Every new element is the evolution of an idea; every new element is born from condensed knowledge, from the only and the same One.

Every particle is born with the help of emptiness; every form of life and Everything from it is life; there is no life beyond it. Its vibrations make multitudes; there is no organic or nonorganic matter. These are our interpretations and linguistic definitions. Everything that is—it is life; Everything beyond it is nothing. The World is the imagined *Perpetuum mobile*, making Everything from "nothing" by "nothing."

Nothing is its food,
Entire knowledge is its energy,
Entire conscience and history;
Its energy saw Everything,
Past and future;
Although a multitude, it does not go anywhere;
There is nothing beyond it.
The way implies the other point
And there is no other value or another point beyond it;
        It, itself, is the Way,
Starting point and destination,
        Target to it.

While flying into itself, it gives birth to new elements;
The endless conversation begins,
Whispering of stars;
Distances give birth to the conversation,
Distances give birth to the giggles of light,
The entire emptiness becomes alive
And starts to resemble space.

RELATIONS

In an absolute vacuum, no space and time exist as actualities except as potential. If there is no world or universe, there is no space and time in our interpretation of these words and concepts. But, if there is no space, there is no world and no time.

The main feature of an absolute vacuum is that it is nonexistent nothingness. But, if we put it like this, it may sound that if nothingness is nonexistent, it exists. If nothingness is nonexistent, it means that there is no nothingness. No world or universe can be made without nothingness because there is no space without it. In actuality, space is nothingness. Without nothingness, which is nonexistence, somethingness or existence would not be possible.

If we analyze the previous paragraph, we come to some strange or not-so-strange conclusions, knowing that language is not the perfect instrument for expressing ideas but is still the best currently available to humans. If nothingness is nonexistent, it means that there is no nothingness. Yet, if there is nothingness, it should mean that nothingness exists. We may think this kind of reasoning is pure scholasticism or nonsense, but in reality, linguistic and transcendental are often in collision. The most abstract and profound ideas often escape or try to escape language formulations. The "weakness" of language becomes more apparent when facing Reality's formidable immensity and depth, especially the metaphysical "side" of Reality.

The main and only primordial relationship is between the Being and the Nonbeing. The Being is immaterial and eternal;

nothingness is eternal, too. In their primordial state, the Being and the Nonbeing become the same. Eternity means always and never, while infinity means everywhere and nowhere.

The Supreme Force, uniting the Being and the Nonbeing, is the Absolute.

The Being is finite and infinite. It is finite while contained within itself. When it is within itself as One, it becomes its opposite—the Nonbeing.

When the Nonbeing is unaffected by the Being, it is a spaceless space, a spotless spot, and finite yet infinite as potential. That is nowhere with the potential to be everywhere. The primordial "space" is absolutely "finite," enabling the Being to inhabit or envelope it. Nonbeing is the exit for the Being because, without nothingness, there is no space. This Nonbeing is infinite in its potential.

Absolute, without the world or universe, is finite. When the Being is equal to the Nonbeing, the Absolute becomes the same—the Nonbeing. The World-Universe is possible only through the active relationship between the Being and the Nonbeing. The absolute potential of the Being and the Nonbeing is initiated and becomes the source of space and time (spacetime), as we understand it, and is the source of infinity as a never-ending potential. Absolute finiteness is infinity because both absolute finiteness and infinity are nothing.

The Absolute is a finite Being containing infinity as a potential. Without nothingness, there is no infinity. Nothingness is nonexistent and, therefore, the most finite thing. Without nothingness, there is no space. Without space, there is no time.

Eternity means Now. Everything contained in one pointless point is an eternal Now. Now, it is Nothing.

Infinity means Nowhere. Everything contained in one spotless "spot" is Infinity. Infinity is Nothing.

If we imagine only our Universe as space in the ocean of nothingness, all the physical, natural forces will not affect the rest of nothingness at a far enough distance from our universe. All this nothingness would just be nothingness.

If we now imagine one more universe, as ours is, there would be some impact of one on the other if they were close enough, and there would be a relationship between them so we could measure distance and have two points of reference.

If we now make a giant leap of imagination, we can imagine a macro universe containing a similar number of universes as there are galaxies in ours; let's say 100,000—1.000,000 or trillion universes, this universe would still have our universe and one million or a trillion more. This universe, at least one million times bigger than ours, would not affect the nothingness much more than a universe one million (or trillion) times smaller. The impact of this universe would be one million times larger and contaminate nothingness one million times more, but still, nothingness would be largely unaffected. The same would happen if we now imagine a universe a million times bigger than the previous one and if we continue to imagine a macro macro universe or a googolplex universe. Even a universe the size of 10^{googol} of our own would not affect the rest of nothingness much more than ours affects it. Beyond the contamination of nothingness produced by a universe with the size of Googolplex larger than our own, the rest of nothingness would stay the same and unaffected.

If all existences or universes were created again and again, along with the new ones, the rest of nothingness would still be equally unaffected and capable of accepting or inhabiting the new ones.

Even if we imagine the googolplex googolplex googolplex size of anything, nothingness would still be able to accept or inhabit Everything. There is no number large enough to reach the end of nothingness. Infinity is simply endless.

All these universes or an Omniverse would be space in the sense we understand it. But space would not be possible without nothingness. Regardless of the curvature of space (and time) or any imaginable physical law, the underlying reality of space would still be nothingness. Space, as we understand it, would not be possible without it.

Aristotle was right in saying that the void is nothing and, therefore, nonexistent. Still, he did not wholly grasp that without this nothing, something, as a physical reality, would be impossible, and there would be no existence. The underlying reality of the Being is the Nonbeing, and existence's underlying reality is nonexistence.

Without Nothing, there is no Everything. Nothingness is the air of Somethingness. Void is nothing, but without nothing, there would be no space, and without space, there would be no "physical world," no time, as we understand it, and no life.

Absolute Void

Absolute void is not and cannot be curved. Without a void, there is no space. Then, what curves if, without the void, there is no space? The curvature of space is the curvature of the immaterial Being that appeared in the void, or enveloped void, and started transforming into something we consider and perceive as physical reality, although it is not physical. Based on our idea of the primary, secondary, and tertiary qualities[6] only as modes of one base quality, we are conditioned, based on our apparatus of perception, to experience and feel the physical world, including curvature of space, as if it were physical, although it is not. The result is the same as if it were physical. The world is the life of the Universal Mind in action.

---

[6] Explained in Chapter 2: "Mind and Matter and Mind-Body Problem," "Programming of Matter," "Modes and Qualities of Reality," and "Degrees of Qualities."

What is Nothing?

When we use the word “is,” that usually means what follows must be something. But if what follows is nothing, this means that nothing does not exist. If nothing does not exist, then the question is why we would even bother saying that. When something is, it usually means that it exists. How can something be anything, even nothing, if it does not exist? So, nothing is only a word, and we can use the “is” before the word nothing in that sense. It still does not mean that the real nothing, of which the word nothing is only a verbal representation, a written sign, exists.

Without nothing, there can be no largest nor smallest of anything material.

What is Something?

Something should be the opposite of nothing. But is this strictly true? If something is the opposite of nothing, then something is the thing of absolute density. Then again, what is absolute density? Is absolute density possible? Absolute Being would be the Being of absolute density. Something with absolute density would lose any dimension and would be non-dimensional. Something with absolute density would have to expel space from itself. Something with absolute density would have to be one because there can be no absolute density if there is any polarity or plurality. Everything must be One squeezed to itself in the primordial singularity. This singularity is **0** nothing. Big Bang comes out of the Zero point of the Absolute.

At the point of their absoluteness, something and nothing become the same. Something with absolute density is equal to nothing without any density. The only way for something to become alive is in the dance with the Nothing. The dance of the Something and the Nothing is the Source of life and the ultimate Source of the Universe.

The material world cannot be infinite. That is a *contradictio in adjecto.* Nothingness cannot be wholly full on the macro or micro level. Material something must have an end at some point, on the macro or micro level.

The Difference between Zero and Nothing

Even in the primordial "form," there is a difference between the Being and the Nonbeing.

Even if the Being is asleep and inactive, in the primordial form, it is still something as a potential.

Although Zero is nothing in a way, Zero is not absolute nothing as real nothing is.

At this absolute "point," the Being is Zero, and Nothing is just Nothing.

Zero is not nothing. Zero is the point between nothing and something (Everything). Zero is a tunnel, a passage, a bridge, a wormhole between the Being and the Nonbeing, infinity and finiteness, eternity and time, existence and nonexistence.

Although zero has the potential capacity for infinity and eternity, it is still the end point of the Being and the Nonbeing when they meet.

The first point, the appearance point, of coming into the material Being is Zero.

Zero is the last point, the disappearance point, of coming out of the material Being.

The zero point is the point of absolute density.

Everything comes into material existence through it.

Everything comes out of a material existence through it.

The "point" where the primordial Being and Nonbeing meet to create is the zero point of creation.

Nothing is just nothing.

Nothingness (emptiness), or absolute void, is Nothing.

At its absolute point, the Primordial Being, the Ultimate Source (God) beyond creation and creating, when it is almost equal to nothingness, is, actually, Zero.

Zero is the "point" where material and immaterial meet.

The last possible "point" of "physical energy" or "matter," in any form, on the micro or macro level, is the Zero "point." Beyond this Zero "point" is nothingness.

SUBSTANTIVALISM AND RELATIONALISM

Substantivalists believe that space and time are absolute categories. Relationalists believe that space and time are contingent, dependent upon relations between some types of entities or objects. The main substantivalist, or absolutist, is Sir Isaac Newton (1642—1726/27), and the main relationalist is Gottfried Wilhelm (von) Leibniz (1646—1716). Newton claimed that space and time are absolutes that do not depend on anything outside them. Leibnitz contended that there is no absolute space and time but only interconnected objects that create spacetime relations.

Instead of going into details relating to these two concepts, I would reiterate my statements throughout this book, which correspond more to Leibnitz's idea than Newton's. If we imagined an absolute void without any material object, that would be nothingness. According to Newton, that would not be nothing but absolute space and time, to simplify the idea. I believe there can be no space without this primordial void or nothingness. The primordial nothingness is absolute in its nothingness but is also absolute in its potential. This absolute potential of nothingness relates to the potential for spacetime relations based on the reality formed after the Big Bang expansion and already established plurality enabling relations among objects.

In this sense, there are no spacetime entities as absolutes with their own existences but only infinite and eternal nothingness devoid of any qualities. Nothingness, or absolute void, becomes spacetime only when somethingness interacts with it to create spacetime. The created spacetime envelopes within itself all the zones of nothingness it deals with. Since nothingness is infinite, it hides an absolute potential for space. Since absolute time is eternal, both time and space can only be finite within the spacetime continuum of the "material" world. Nothing that exists in the world that we consider physical can be infinite or eternal. Everything that exists within a physical realm has a spatial or temporal limit.

Newton was right that there is something absolute, but this absolute was not space as he had seen it or time as he had seen it, but nothingness as an absolute potential for space and time. But, for space and time to make any sense, relations are essential, for without relations, there would be no motion, and without motion, there is no time.

The underlying reality and basis for space is nothingness. Without nothingness, there would be no space and no time. Nothingness hides the secret of infinity and eternity. Infinity and eternity exist as potential and not as actuality.

Italian scientist of Croatian descent, Francesco Patrizi (Franciscus Patricius; Croatian: Franjo Petriš, 1529—1597), a century before Newton and Leibnitz, investigated the question of space and wondered, "[i]f space is nothing, does it exist?"[7]

---

[7] If space is nothing, we ask in vain if space exists. If space does not exist, we likewise ask in vain if space is something else. It is a notion common to all people that space is and that it seems to be something, such that it was given the following or similar names: dimension, distance, interval, diastasis [interval], diastema [extension]."

– *Francesco Patrizi*

"Si spacium nihil est, frustra quaeratur, an spacium sit. Si verò spacium non sit, frustra itidem quaeretur, an spacium sit aliquid. Communis quaedam omnium hominum notio, spacium, et esse, et aliquid esse videtur voluisse, cum nomina haec, vel talia formaret, Dimensio, Distantia, Inveruallum, Intercapedo, Spacium, Diastasis, Diastema."

– 1587, 2v.

According to Newton, space is not nothing, and it exists. Then Patrizi asks, “If space does not exist … is space something else?” According to Newton, space exists and is something else.

SPACE IN THE UNIVERSE

EINSTEIN'S SPACE

The space we explored and elaborated on before is not the same as we experience in the existing Universe, as we "see" it. In the "space," before the Universe, as we know it, there is no point of reference. A point of reference exists only if there is a relationship between something and something else, even if that relationship is only between two things, two entities.

We will call the space we experience Einstein's space. Einstein's space is "impregnated," programmed, or shall we dare to say, "contaminated" by the awakening, "explosion" of the Absolute. This explosion is the dispersion of Nothingness into the Being, not the "explosion" of the Being into Nothingness. The "space" created this way is not the clean "space" we described as an absolute vacuum, emptiness, or nothingness.

If we talk about the Primordial Being and Nonbeing, we cannot talk about space from Einstein's physics point of view. In the primordial "space," the Being does not possess any material properties; the same applies to the Nonbeing or nothingness. In such a state, there is no space as we understand it. Every point is the same point. Every moment is the same moment. In such a state, the vastness of "space" and "time" are the same point and moment. This is the infinity of the finite, compressed infinity of the Absolute Being.

The Infinity and Eternity outside the realm of the “material” Universe are the Infinity of Eternity and Eternity of Infinity enslaved beyond space and time. That which gives space and “time” to the World is beyond the spacetime continuum. The Universe is the manifestation of the Absolute Being. Absolute Being is spaceless and timeless. It is everywhere and nowhere at the same time. It encapsulates all the space and time, yet it is outside space and time. Absolute is spaceless space and timeless time. Absolute is Everything and nothing at the same time.

CURVATURE OF SPACE

Emptiness, an absolute vacuum, cannot be curved, yet only emptiness makes space. Only matter, our idea of matter, can be curved; its curving is not a condensation of energy but the dilution of the energy of compressed "knowledge," "mind," idea," "thought," or "spirit." Only this condensed essence, "immaterial substance," can be curved in its expansion or "extension." Neither the "real space" (absolute vacuum) has an actual dimension, nor does the universal essence, the "ultimate substance," have it. The "immaterial substance," Universal Mind, is the only authentic energy of the Being, existing in its knowledge and "memory."

Space, in its essence, does not exist and, therefore, cannot be curved. What exists is only an emanation of the Supreme Being transformed into reality. We measure the expression of the Absolute and not the absolute itself. We already elaborated on the manifestation of the Absolute as we see it. Therefore, the curvature of space is a curvature of the convention, as presented to our senses, the curvature of the emanation and manifestation of the Absolute Mind in the World.

Reaching the ultimate laws of physics requires distancing from the paradigms of science and the paradigm of how a human brain usually thinks or functions. To understand the ultimate laws of physics, the human mind must adjust to the requirements or logic of phenomena that seem paradoxical and absurd to us as humans.

A "collision" between the Being and Nonbeing gives birth to space. Their merger is annihilation or, to use Heidegger's term, nihilation and birth simultaneously. The dimension is born from this merger—third, fourth, or many more. This question is not as important as we think unless we believe that senses are the measure of the Universe and God and that the dimensions are what we think they are: third, fourth, twelfth, or countless numbers. Nevertheless, we see dimensions. What constitutes beauty is not so much in the discovery as in the voyage itself, not in the final arrival to the ultimate conclusion, but in the magic of being, the magic of existing.

Every dimension looks at itself through the eyes of other shapes in the Universe. Only in the senses, and through the senses, the joy is born and felt. The Universe's uppermost beauty, depth, knowledge, and potential shine on surfaces. The Universe puts on and changes many clothes, colors, shapes, and scents. Light is the robe of the Universe and is more important than its depth. Its whole nature shines from its face; its mystery seduces us with the smile of beauty; its depth is in the service of beauty. The most profound depth is sad and lonely at the bottom of itself and does not need knowledge, for it is in and of itself omnipotent knowledge and truth. That's why the Being, in a creative dance with Emptiness, embraces her,

To create a smile on her face
Which becomes space
Giving birth to distances among shapes.
Without shapes, there are no distances or Paths,
Without the Paths, there is no purpose
For the Ultimate Path is the purpose,
The Path walks the Way to the Harbor.
Space is the Ultimate Curved Path,
On the Way from Oneness into Plurality
And from Plurality back to Oneness.
Plurality is the curved smile of Oneness
In myriad colors and shapes

On countless paths of the divided One
Toward the Ultimate Home of the One
Undivided and uncurved Self.

A journey itself is beauty: space, journey, is the hope and joy of the world. Nothingness is the most important "dimension" of the Absolute or God; he doesn't have dimensions without emptiness. Without Nothing, he is Nothing. Without the Nothing, there is no world and ultimately no space; if there is no space, there is nothing to be curved. However, if there is something, that something can exist as the world, only with the Nothing or in the Nothing or the Nothing can exist within it. Nothingness cannot be curved; only the Being can have curvature. The curvature of the Being is only its "physical" appearance and not the actual curvature of space, which is nothing and cannot be curved. As manifested in the world, the Being is many and One simultaneously. It is divided into a multitude yet stays One. There is only one Universe containing plurality—micro shapes, micro-universes, particles, waves, whatever we call all the elements of existence.

What causes the illusion of the curvature of space is the limitation of language, thought, understanding, and the application of our limited language and beliefs to reality. We cannot change reality, but we can try to accommodate our thoughts and language in what appears to be a reality. Since the Nothing is "not" space, we are sure we can legitimately say that space can be curved. Nevertheless, we must first answer what space is. If we cannot precisely answer what space is, we cannot speculate that an absolute vacuum is not space. In other words, we cannot base our thoughts and language on consequences and conclusions rather than causes and premises.

The main question is if there can be any space without space, although this sounds absurd, for it is evident that there can be no space without space. However, if there can be no space without space, how can there be the curvature of space if there is no space?

Then, we may answer that there is space and engage in circular reasoning. On a superficial level, some "obvious axioms" lead us to accept that there is space as it is without questioning how that is, how it is possible, and what creates this space. Then, we may answer that, based on laws of physics, nature works in such ways and that there are four main forces, and that based on all our knowledge and theories confirmed by experiments, we conclude and state, based on a "fact," that there is a curvature of space.

We are still too far from the true definition of space and a true definition of fact. To prove that there is a curvature of space, we must undeniably prove and show that we know and understand what space is and that we know and understand what a fact is philosophically, linguistically, and physically. Consequently, we may say we have considered and know all this, but the "fact" remains the "fact." It is not whether we say we understand all this and have considered all that but if we have genuinely and undeniably proved and understood what we declare.

In continuation, we may argue that since nothing is just nothing, it cannot be space, and since it is not space, space is, in other words, what we see or think space is. However, what we see or think about space does not mean we have proved that what we see as space is real space. What is curved is only the Being, presented to us as something we call matter and energy, not space. Although Being envelopes space or is enveloped by space, it does not mean that space curves, only the Being. Since space is undetectable, passive, and nonexistent except concerning the Being, it does not contain any property or feature and, therefore, cannot be curved. There must be something to be curved.

Nevertheless, without the Nothing, nothing can exist in the world as we see it. Nothing is the medium of the Being. Without the Nonbeing, there is no Being. This coexistence and symbiosis produce something we may call, at best, an illusion of "curvature of space" since actual space remains uncurved and cannot be curved.

We still can insist that space is what we say because nothing, or an absolute vacuum, is not space. Then, we must prove how space can exist without this nothing or absolute vacuum. But we cannot prove that space is possible without nothing. Since we cannot prove that there is space or curvature of space without nothing, we will establish the opposite: without the absolute vacuum, the Nothing (void, emptiness), there is no space. When we become convinced that there can be no space without the Nothing or an absolute vacuum, we must answer precisely what space is and how it becomes space. The only actual space is nothing or an absolute vacuum; we have already stated that this space cannot be curved.

We may still insist that space, although mainly emptiness, is completely "contaminated" by all the forces in the Universe—particles, waves, and so on. Still, based on this deduction or inference, we cannot prove that what is curved is space since nothing remains nothing and cannot be transformed into something based on the laws of physics. If the Nothing does not convert into something, the curvature of space is impossible. To prove the curvature of nothing (space), we must prove that nothing can transform itself into something based on the laws of physics.

Suppose we further insist that this is only a problem of linguistics and philosophy and not physics since we stated that space is how we define it and not as it is. This reasoning would be insufficient because we must first prove that actual space is what it is and not what we say it is to fit our arguments. There can be no answers to the most critical questions of contemporary science, physics, philosophy, and even religions if we are not as precise as possible, linguistically, experimentally, or in any other way. These questions appear to be self-evident, self-explanatory truths and axioms, yet they are neither self-evident nor clear and precise to reflect the actual underlying reality. If we are not as precise as possible, we create theories and paradigms presented as facts, although the starting premises are undefined and unanswered. However, if we do not answer the starting causes and premises, we cannot adequately describe the laws of nature, nor can we understand them.

Regardless of how sure we are about space and the curvature of space, we still cannot claim we are correct without describing the nature of nothing. Without the Nothing, there can be no space. There will be an immediate argument, though, that the Nothing has no nature and, therefore, there is nothing to describe. The answer is that its passivity and lack of properties are its most potent "property" because they enable creation and existence. Without the Nothing or void, there is no creation and no existence.

We converted nothing into something by our thoughts and language, using deduction or inference, and concluded that space is the consequence of this thought process, not the actual process. We applied our definite language to our indefinite "understanding," ideas, or reasoning to prove the "fact." However, the fact is or is not, regardless of our ideas, language, or reasoning.

We must use the opposite and apply language to the facts rather than our understanding. Our understanding is limited, and facts are impersonal and independent of our knowledge. We cannot falsify the language to fit and understand the facts better. We cannot change facts or affect them with our ideas, but facts are verifiable up to a point. There are facts beyond the verifiable point by humans since human beings are limited. Still, language is verifiable, and our theories are both falsifiable and verifiable to a large extent.

Since the immaterial Being envelopes the Nothing, this nothing, in "cooperation" with the Universal Being (Mind), becomes space as we experience and describe it. The "fabric" of this created space, with the help of nothingness, is curved. But all this is the product of the transformation of the Being into its different forms, modes, and interdependent qualities of reality. The "material" world is only a symphony of "materialized" qualities of the Universal Being," not matter *per se* because matter *per se* does not exist. What we see as space is a "materialized" program of the

Universal Mind. What appears to us as dimensions is the underlying nothingness holding the illusion of Reality, making it appear material. The Primordial Primary Quality is the Primary Ultimate Force, or Source, that powers all we see, experience, and measure. Everything is related to Everything else and is affected and conditioned by Everything else. Everything within the Universe is a message, information, and code to everything else. Energy and matter are the messages of the Universal Mind sent into nothingness to fertilize it. Relationships and communication among the myriad beings are the life of one organism. All the features of matter we experience are real in the sense that we experience the spacetime continuum, but all that is the result of programming and conditioning rather than energy and matter as physical realities *per se*. Everything was One and became a multitude, yet Everything stayed One on the most basic level. The story of One is the story of All. Every sense, every pain, and everything we feel are the messages of existence, messages of the Universal Mind in action, interconnecting the Web of the Universe into One Family. Absolute is the Ultimate uniting force of Everything. To be one and only is death. That's why creating is needed. Without creating, there is no life. The Universe is the life of the Absolute. Something and Nothing are the Father and the Mother of the World. From One Absolute, there is an almost endless family dispersed through space in search of life and meaning, which is what we call existence.

Curvature of Information

Primordial primary quality informs all reality, interconnecting and providing everything with adequate apparatuses to operate within the secondary (formerly primary) and tertiary (previously secondary) qualities. Space, with the objects in it, is the information we receive as presented to our senses. For instance, we can measure the length, depth, and width of anything, thinking that we measured the actual space. However, we measured nothingness enveloped by "information," which creates space or reality presented to our senses. The underlying reality of everything always stays the same: nothingness in the Web of Information. The "volume" of anything is nothingness. Whenever we measure something, we measure the message, information, or appearance (illusion) of space enveloping nothingness. Nothingness is the actual "volume" (space) of anything, and the appearance is immaterial information occupying it. All the void "touched" by the Primordial Being and its "Force" is "contaminated" and cannot be treated as pure nothingness.

The curvature of space is one of the consequences of this "programming." We live in the Web of Information, which we experience as the "material" Universe. The Universe is the Web of Information. The curvature of space is, in fact, the curvature of information, not of space or matter. This reality, which is information, provides us with data about its qualities, properties, and features that we can analyze and measure. However, all that is an "illusion," actually an objective reality, different from the reality we thought we lived in. Reality is a convention. Although our experience of reality may be the same despite the new knowledge, our understanding and concepts will most likely shift in ways unimagined before. The new understanding of reality requires and may cause a major paradigm shift, perhaps the largest one in the history of humankind. Unless obstructed by some natural cataclysmic events or wars, this paradigm shift may result in a new renaissance in society, science, and the arts like never before.

TIME

The whole past, the present, and the future are in the world as potential. Without space, there is no growth and time. Growth and development are possible only in the multitude, which implies distances and is not accidental. Even if the beginning was accidental, and it was not, growth and development are not. A compressed idea, the primordial Universal Mind, contains the Way and every dimension. Time is the measure of distance; time beats from its heart and records the motion of its bloodstream; time turns the pages of the Way and epitomizes the unmistakable memory of the Being-God. Time is the Being's device with which it deceives the emptiness within itself. Time records and glorifies the glory of the Being-God.

Since both past and present are in him, he has already walked his way; he will not arrive at any other time or have more or less time. He will return to the same "place" simultaneously, beyond time. He swims outside time and seeks and offers the purpose through the artifice of time. No particle is aware of its most profound memory. However, every particle rushes through time in its unmistakable walk, and every mistake is part of this unmistakable walk, part of its perfection.

Life is impossible without a "mistake." Perfection, without an error, is death. Nothingness must be perfect because nothingness is incapable of a mistake. The only real "perfection" is life. Life is the price for a mistake and deceit.

Life has no meaning without the past and future; the present, without the past and the future, is the frozen moment. We all fight for the present, but if anything exists in time, that is the past and the future. Is it possible to measure the present? How long does it last? The present second is not the present: before we think about it, it becomes the future; when we think about it, it becomes the past, and the future becomes the past in the exact second. (This is a good argument about particles and waves since we cannot determine a particle's exact position and momentum.) What time is there in the present? Only the Absolute is in the absolute present because the absolute present is timeless and spaceless. The present is eternity, and eternity is nothing.

If we divide the second, every fraction becomes the past faster than the speed of thought and every fraction of every fraction. We would need to step down to the Zero Point of time to measure the present; we would need to divide the smallest number by the infinite number of numbers and look for the past in the infinity of the second. The whole matter is smaller than the numberless "number" of infinity. Infinity is endless.

Even endless knowledge of the Universal Mind cannot sustain infinity as actuality. The Universal Mind cannot transform into an absolute plurality of infinity. Its purpose is to transform from one into many and to exist meaningfully. The meaning and the purpose of the Universal Mind are saved based on this limit imposed by something without any limit, the infinity.

Real endlessness, absolute multitude, and variance imply an accident or the impossibility of an accident, which boils down to the same. Absolute multitude and endlessness can exist only in emptiness. The Home of endlessness and the multitude is Zero. We have to reach the bottom of a second, the bottom of time, to pass through the whole multitude and "infinity" to catch the present. Nevertheless, it is impossible to catch the present because it is Zero. Over time, the future becomes the past at the edge of Zero. The present is the annihilation of life; the present does not exist, but Everything is the presence of an organism that remembers and breathes according to its program—the present is the eternity it tries to avoid. All the rest implies some motion or illusion; it resembles eternity because it has eternal life.

Zero is the only thing endlessness and multitude do not want to possess. Zero is the only eternity—eternal sleep, but endlessness, the Absolute, wants motion and life. Eternity is possible only beyond perpetuity. Duration implies the number, endlessness, and space. Only Zero enslaves every number, every endlessness, and space; only Zero enslaves (captures) perpetuity (duration). Only enslaved perpetuity can be an eternity, and eternity can only exist (be) in the present. Every past is measurable and reducible to a number, and so is every future; only the present has no number. The only number of the present is Zero, and Zero is the "end of life." Therefore, there is no life in the present, although it is the only life if we think from the point of view of life.

Zero is the enslaved time, enslaved past, and future in the present of Zero. Zero is the end of recording: the time that contains all time and an absolute multitude within itself. Time, an eternity, is smaller than a moment, than a second, than a fraction of a second; it is not measurable. Time does not exist. Eternity does not have measure; its only measure is Zero, into which it goes to sleep. That is why an instant is longer than an eternity: an instant has a duration.

RELATIONSHIP BETWEEN SPACE AND TIME

Zero enslaves space and time when a compressed idea, the Universal Mind, explodes into itself through emptiness; when a compressed idea, the mind converts and irrupts into countless ideas (messages). Creation of space is the classification and ordering of ideas (information), not division or motion but development, birth, and happening, the life of an idea, and development of the musical scale. Space is the trembling of emptiness: the life of a mind that listens to itself.

There is no time and space. Motion is a Sisyphus' stone, and space is emptiness, which is nothing. Zero enslaves the space within itself, leaving emptiness without space out of itself. Zero frees the space from the condensed idea, offering the emptiness to spread within the compressed mind. The enslavement of emptiness is the birth of space. Freedom of emptiness is her slavery. Emptiness becomes the slave and the master; obeying God, the vacuum gives him birth, in a way, by providing space for plurality, but simultaneously enlivens itself.

God is Sisyphus, and existence (life) is the stone. The accomplishment of purpose is the saddest point. Sisyphus is happiest without the total accomplishment of the purpose, while he persistently works and aims at the top. The accomplishment of the purpose is Sisyphus' fall, so he forgets his achievement quickly. Voyage contains his whole hope and beauty; the voyage is the purpose of the stone—purpose to itself alone. He always arrives in the same place but chooses voyage instead of the target. The purpose and target were always there. Voyage is always new because it is fed from oblivion by unaware memory, with infinite possibilities stemming from the nature of the absolute through "free" will and unincidental accidents or errors. Voyage is the ultimate goal or purpose because it hides an infinite multitude within itself; the infinity of the finite through endless possibilities. The beauty of infinity is always new; the voyage is the sum of happenings on the scale. Every experience or adventure is new. Beauty shines from experience.

Only the possibility of a journey creates space because space, before and after, becomes (is) the same; only the Way on a journey creates time, and the idea of the goal gives birth to time. Time does not measure time but the length of a passed road. The journey is space since nothing journeys (travels) from one to another or apart from one another. Nothing is ever separated; rather, all travel with all; all looks for itself within and in the other, and every path is the idea of the Being that finds itself on its journey because it is the Way.

The World is the Way because there is nothing outside, and through nothing, there is no way: The World travels through itself and within itself. Time is the measure of its Way, and since the world is the Way, the Way travels, not the world. The Way is the only measure, the only time and space. World-God is the Way.

CHAPTER 4

WHERE IS THE END OF ANYTHING?

Only (the) Nothing can be infinite because the Nothing has no beginning, end, or limit. Nothing can fill in Nothingness. Only Something with its properties (and features) can enrich itself or accept features and properties, yet Nothingness is without any features or properties. The featureless "nature" of Nothingness is its uppermost value and "asset"' because the Being, in the pursuit of creation and creating (recreating), needs precisely this featureless feature of Nothingness, its most potent Force, which is the lack of any force, the forceless force.

If we try to imagine the fullness of nothingness, we will realize it is theoretically impossible. A void cannot be partially full (filled in) to be full but would have to be full in its totality to be full indeed, which is impossible.

Nothingness would cease to be nothing and become the opposite if nothingness were full. That is impossible (although nothing and something, on the metaphysical or symbolic level, can be the same if something is asleep).

The only way for interaction and mutual influence between nothing and something is through the activity of the One we have chosen to call Something. This Something "colors" nothingness with its own colors. This Something envelopes Nothingness. Regardless of how strange it sounds, Nothingness is never full, not even a bit. The Immaterial Being envelops the Nothingness in the form of a "material" Being, the Universe, yet all the happenings of the Something (the Being, Universe) are immaterial "forces" transformed from its primordial stage based on the principles of interaction between the primary, secondary and tertiary qualities.

The underlying reality is the illusion of space, energy, and matter. All we see is nothingness, untouched by the happening of the Being. Since nothingness cannot be curved, what we experience as the curvature of space is the curvature of the Being presented to us as energy and matter, yet immaterial. But, "immaterial substance" (Universal Mind), although immaterial, is still something. This something has the potential to appear as material in the modalities of transformation of the Universal Mind. That is the creative potential of the Universal Mind to secure its creation and, more importantly, the meaning stemming from the creative power of the Ultimate Source.

Back and forward or in any direction, we can measure no space except nothingness. Within the coordinates of the Being, we measure the "space," but that which we call space is only the appearance of the immaterial Being presented to us, or sensed by us, as material. Yet, the only "volume" of this "space" is Nothingness.

What we experience as Space is the most beautiful "illusion" filled with Nothingness. When we say Nothingness cannot be partially or fully full, we mean the opposite: only the Being may be full, not only partially but wholly full. The wholeness, fullness of the Being, comes from Nothingness. Without Nothing, Something is without the volume. Without volume, there is no "space."

Nothing can be in the Nothing. Nothing can be only in the Something. Without the Nothing in the Being, there can be no Being, the Universe, in the sense we experience the World, and there can be no space. Whole Space is Nothingness contaminated by the "illusion" of the Being transformed into Something we call Universe.

The biggest paradox of Nothingness is that its nonexistence creates infinity. The lack of any property makes the infinity of nothingness. Infinity is nowhere. If the point of infinity were identifiable, its end would be known, which is impossible.

Neither Something (the Being) nor Nothing (the Nonbeing) have an end or could have an end. That which has always existed does not have a beginning or end, and that which has never existed cannot have a beginning or end.

Nothingness, or its possibility, derives from the world with its definite properties, which we experience as dimensions “shaped” by our senses, feelings, and thoughts. Our experience is our ultimate reality. Without this reality of plurality, the ultimate truth of the Being (Oneness) is less natural and meaningful.

Dimension, or something that has an extent, cannot be infinite like previously described nothingness. If this were possible, there would be two “infinities”—the infinity of the “Absolute” (God) and the Infinity of Nothingness. These two principles would not function in correlation and influence one another (servitude and dominance) as active and passive principles. As two absolute infinities, these principles would function as the active (or passive, with no difference) principle in both cases. Such a relationship would only lead to mutual annihilation, not coexistence and influence. These worlds would become so different that this absolute difference would merge and convert them into the same principle, regardless of how paradoxical this may sound. But this would be the death of both, and death is not possible in either case.

Since it is possible to imagine either one or the other concept theoretically, it is impossible to defend this “concept” even theoretically or as a hypothesis. Infinity is only a potential, not an actuality. The Being and Nothingness are “finite” as actualities and infinite as potential. Nothingness cannot have any potential *per se,* yet the Being cannot exercise its potential without the Nothingness with no potential. The lack of any potential of the One feeds the potential of the Other. The power of Nothingness is equally “forceful” as the power of the Being. Without Nothingness's “forceless” force, there would be no force of the Being or the possibility to exercise its power in creating the World (the Universe). Absolute passivity of Nonbeing is equally vital as the activity or dominance of the Being.

NONBEING (NOTHINGNESS)

Only nothing is infinite in its endlessness.
Its endlessness is its end.
(The impossibility of a beginning and end is the secret of infinity [eternity].)
The endlessness of Nothingness is not real infinity.
(Infinity of Nothingness is Zero.)
Nothingness is finite because it is reduced to Zero anytime and anywhere. Anywhere is nowhere.
Nothingness must be endless if there is nothing to reduce to Something.
The biggest limitation of Nothingness is that it is limitless.
The lack of limit is the limit of Nothingness; its infinity reduces it to nothingness.
Nothingness is neither finite nor infinite.
The nothingness of itself and for itself is absolute.
(God of itself and for itself is absolute.)
Nothing (Nothingness) and Something (God) can be absolute because if they don't disturb one another, their separateness reduces them to nothing. They can be absolute because, as separate absolutes, they are the same and one.
Since Nonbeing (emptiness, absolute vacuum, nothingness) is nothing, it is passive and compatible with the Being; it is the perfect medium for the finite Being (infinite in its potential). Nonbeing offers no limits to the Being, which allows for its potential development.

THE BEING

Being (God) is finite.
Only a finite being can exist.
That which exists must be finite.
That which does not exist is "infinite" potential.
The potential is potential existence.
That which may exist already exists in a different form.
The infinite potential is the ability of the Being always to be new.
Nothingness exists in the form of nonexistence.
Existence of nonexistence exists.
Infinity of the finite is the secret of existence.
The very finiteness of God leads to infinity.
The finiteness of God is absolute.
(The Finiteness of God is Zero.)
The finitude of God hides infinity (potential).

RELATION BETWEEN THE BEING AND NONBEING

Nonbeing plays almost an equally important role as Being. By lacking form, the Nonbeing offers life to the form. By lacking limits, the Nonebeing provides the opportunity and space to the Being (with limits).

Being is finite, and only something "nonexistent" can offer infinity to the finitude of the Being. Being and Nonbeing are the alpha and omega of existence. They negate each other and support each other. They fight with each other and make love to each other. The result is a Living Being, a Living World. The lack of one is the death of the other. Without the absoluteness of nothingness, there is no absoluteness of the Being. The absoluteness of one is equal to the absoluteness of the other. The Being and Nonbeing, without the World or plurality, become one, and this Oneness is the negation of both or the transformation of both into nothingness without the attribute of absoluteness (the Nothing cannot be anything else except nothing).

Conditional absoluteness of the Being is, at the same time, proof of its relativity and the relativity of its antipode. This very relationship makes them relative. Absolute itself unites the Being and the Nonbeing. Lovemaking of the Being and Nonbeing results in the World's birth. The world is relative because it comprises two "absolute entities"—Being and Nonbeing. It would not be able to exist without the Nonbeing as it would not be able to exist without the Being.

If Being is the "vital energy," then the Nonbeing is its space. Without the Nonbeing, the Being would die from its "weight" and have nowhere to go except farther into itself. But moving farther into itself would be impossible because the Being, in its primordial form, is absolute, implying absolute density and concentration in a static state of nondensity. Absolute density is nondensity.

Nonbeing provides the "air" to the Being to spread its wings and fly "through" nothingness. The finitude of the Being is the source of its infinity through the Nonbeing. This infinity is not measurable because it is not possible in any particular life or any particular circle, but in the infinite number of lives and cycles of the same thing, living the life of constant change, constant annihilation, and rebirths, *ad infinitum*, through and by the infinite number of variations of the same finite and omnipotent Being.

The power of the Being is not in its infinity, which is already "accomplished" in its finitude, but rather in its life. Its task is not to impress itself with its omnipotence and power but to preserve life. Its only purpose is life, and its only glory is manifested and represented by life. Being is a Phoenix and Sisyphus. The Nonbeing is Atlas holding the whole Universe on the "shoulders" of nothingness.

BIG AND SMALL

At the point where the big is so big that it almost touches the "farthest" possible "end," it becomes static. The state of absolute speed is the state of no motion, the static state. At the absolute speed, the starting point is the ending or destination point. Appearance and disappearance happen at the "point" of absolute speed. The creation of energy and matter (the world) and the disappearance of energy and matter happen at the point of absolute speed through zero. The conditional biggest has nowhere to move, nothing to revolve around. The biggest must be near the point of Zero. It has nowhere to go except into the smaller. The life of the big is the life of the small. At the "point" of Zero, distances disappear because everything disappears. Manifestation of the Universal Mind, in its physical form, returns to itself in its primordial, immaterial form beyond space and time.

We cannot be sure how many universes make up the Universe. The Universe we live in, which began based on a theory of the Big Bang almost 14 billion years ago, is only one of the many, comprising a much bigger Universe than we think. We believe the energy of the Omniverse can explain the dark energy for which we cannot find the source or explanation.

Considering this, we cannot be sure if this "ultimate" Universe, Omniverse, ever disappears in its aggregate state or is constantly making and remaking itself through and by the ultimate Force or Engine of Everything—the Universal Mind. The Omniverse is another face of the Absolute, its manifestation in plurality.

Omniverse is the Universal Mind in action, spread out in a limitless number of its potential parts while staying One simultaneously. Then, all parts seek to return to the primordial state of absolute peace and Oneness.

The One always remains the One. The division is appearance. The division and the life or existence of many is a phenomenon. One or Onenennes is Noumenon. Oneness is always present in plurality. Oneness is an underlying reality or Noumenon in action. For example, a universe like ours may have its birth and death, appearance and disappearance, while the universe as a totality, the sum of all, may still exist. In that case, Carlo Rovelli may be right in saying that "there is no such thing as a real void, one that is completely empty"[8] (physicists have a consensus about this). Still, no universe, regardless of its size, can encompass nothingness or emptiness in its totality. All "space" beyond, not confined or affected in any way by any universe, is a real void. We cannot talk about a real void within the Universe, but beyond it, there is such a thing as a real void.

The other question is whether we agree about the terminology or if we need to adjust our language to the reality to be more precise. We would have to define different terms precisely in all their meanings and varieties. We cannot talk about the "real void" if we do not explain what a real void is. Only when we define the terms precisely can we use them freely, knowing that we will be as clear as possible and that others will understand us correctly; otherwise, there will always be room for misunderstanding.

We believe we understand the real void, but do we know exactly what it is? Perhaps we think we know what space is, and based on that knowledge, we believe we understand what void is in the context of space. But what if our idea of void is wrong in the context of space as we know it?

---

[8] Rovelli, Carlo. *Seven Brief Lessons on Physics.* Penguin, 2015.

In the way we understand space, there is no real void, but in the way I described the void in the previous sections of this book, the real void is always present. What we see and experience as space is an illusion. It is impossible to transform a void into energy or matter and curve it. The real void is uncurvable. What is curved is the "immaterial" Universe in the form we experience as space, energy, and matter based on the realities predicated by the Ultimate Primary Quality of Reality and based on secondary and tertiary qualities we have already described.

Real Void is in the space yet unaffected by the made "space." Although it is nothingness, the void provides volume to the Being to create space. Whatever we see and experience as space is a real void, regardless of how strange or absurd it sounds. As we experience it, the appearance of space envelopes void with its information and laws, offering information about the curvature of itself (the Being), not the void, yet the real void is always there and is "indestructible," unchangeable, unaffected, and uncurved.

Nothingness, or real void, is untransformable except as a medium within something else that may appear different. What seems different is not the void but the "space" created by the Being with the help of the real void that still exists within the space as an underlying reality, providing volume to that very space we experience as curved. What is curved is the "information," laws, and structure of the Being manifested as "space," the Universe, not the real void or nothingness itself, without which there is no volume and therefore no space.

The real void is the medium for the Being to "create space" or the appearance of space with the help of the real void. Without the real void, there would be no volume; without volume, there would be no space. Although passively, the Real Void "participates" equally with the Being in creating the Universe. The passivity and nothingness of the void are its biggest "asset" and "force."

Our idea of the birth of the Universe through the Big Bang is only our limited idea about the Universe and not a real explanation of the Universe itself as it may be and not as we think it is. Therefore, we would have to define, in the first place, what a real Universe is. We have to determine whether we can always use our ideas about the Universe as a basis for understanding the Universe. But, if we are not sure we understand, or know with certainty, or at least to a high degree, what a real Universe is, in its totality, known and unknown to us, then we must think about it more hypothetically.

Can we go beyond the existing Universe to establish some scientific truths? However abstract this may sound, we may question relations and influences. For example, if there were other universes anywhere before the Big Bang of our Universe, how could that have affected our Universe, and would these effects be measurable or if there were any? Many other questions relating to impact and influence may indirectly lead to latent potential answers.

Then, we must be sure we understand what creation and recreation are. According to atheists, the world is accidental. But is that possible? I firmly believe that science can prove and will prove that matter can't come into existence just of itself or be always there just like that. It is almost inconceivable that matter, if it did not possess any information enabling it to function, would be able to evolve, in some instances, to the point of awareness of itself and the world around. It is almost impossible that matter, or energy, as such, originated just of themselves or always existed.

When the small is so small that it almost touches the farthest possible end of "infinity," it becomes uncatchable. It is the point that we freely choose to call the "point" of "absolute speed" when the smallest, or near the smallest, when looked at, appears to be a particle and, when not looked at, seems to be a wave. The smallest must be the trembling of "matter" near the point of zero. It has nowhere to go except into the bigger. The life of the small is the life of the big.

Infinity is the tunnel between two zeros, the meeting point of two zeros, which is the same point. From the outside, the Universe is static. Motion is an illusion. The Reality of the whole Being, regardless of how magnificent it is, depends on Nothingness, without which it has nowhere to go. Its life and meaning are based not solely on its power but on the characteristic of infinity to be endless, which empowers the infinity of possibilities.

Absolute is always the same, yet always new. Its sameness is the source of its variety. Its sameness and oneness (singularity) make it omnipotent and rejuvenating. Its infinite age makes it infinitely young. It cannot be born or die, but it can live. The ultimate exit of the Absolute is life. Its meaning is life, not absoluteness. It already possesses absoluteness, but its absoluteness is its biggest enemy if it becomes satisfied with it because it transforms it into nothingness. It must fight against its absoluteness to gain the absolute value of life, which, although relative, provides meaning and purpose.

Without meaning and purpose, power and omnipotence become weakness and impotence, reduced to nothingness. In this sense, the relative is "bigger" than the Absolute. At the highest point, the biggest becomes infinitesimal and meets the smallest at Zero. The Biggest and the Smallest become the same. The meeting point of Zero is where the Absolute exists in its primordial form, equal to itself and its poles. There is nothing to establish the difference between Being and Nonbeing.

Only the hierarchical law of values can establish order and relationships. Only within this system is life possible. Separation creates life. One is longing for another. The game of reacquiring "lost knowledge" starts. The smallest of all worlds, the Absolute itself, becomes the biggest. Although perfect in "design and function," it is "imperfect" (in its actual realization through the world) enough to allow for life. Perfection is only nothingness and the Absolute (in its passive form). Real "perfection" can only be "perfect" if it exists as a personification of the Ultimate Being. The

Universe, or Omniverse, is the personification of a universal and absolute mind of the Being.

To exist, the Being must choose a possibility of chance. Paradoxically, a chance is not accidental but must be almost accidental by nature. Freedom comes from the possibility of a chance. Lack of chance kills freedom. Even variety is a program coming from the ultimate potential of the Being not only for infinity but for freedom, too, in different scenarios of evolutions of different universes. However, without the possibility of a chance, freedom is lost forever. The Absolute Being can even deceive itself through chance and produce an illusion. Without a possibility of chance, life would almost lose its meaning and purpose. There would be no real variety except for the predestined outcome.

If the outcome is predestined and known, then there is no reason for a journey from the point of view of the Absolute Being. The way and the result are already known and accomplished; the road passed before the journey started. The Absolute Being must have an "illusion," precisely as its elements have it, that the journey is unique. And this is not an illusion because there is an infinity of possibilities. Even the Absolute Being cannot reach the end of infinity because there is no end. The Absolute Being will always be able to live in the new Universe, a new world. Even if they are alike, they are always new and unique. No two worlds are the same, but theoretically, repetitions are possible. Through possibilities (potential), this relativity of the Absolute makes the Absolute possible (as life).

How big is the biggest? How small is the smallest? Can the smallest be the biggest and *vice versa*? (The Big Bang, for example, may serve as proof of such transformation.) The biggest is big enough to become the smallest. The small is small enough to become the biggest: the smallest and the biggest meet at their endpoints (edges). The endpoint of the smallest is – Zero. The endpoint of the biggest is – Zero.

Quarks and Waves.

The biggest is equally distant from nothingness as the smallest is. There is equidistance from nothingness to the biggest and the smallest. If we look at an electron, it appears to be a particle and a wave. The notion of a particle as a wave comes from the fact that we cannot measure its position simultaneously when we measure its momentum (velocity). This is Heisenberg's uncertainty principle. If we look at a particle, it is a particle; when we do not look at it, it is a wave. The sheer identification of the particle at the moment we look at it does not mean that we, or instruments, changed its nature but just identified it as such. It is not identified and appears to be a wave when we do not look at it. On this level, being a particle or a wave almost does not make a difference because this is "near" the "point" or "near" the passage from the Ultimate Immaterial Reality to our sensory reality. These are the passages from one sphere into another. At this level of infinitesimal values that instruments cannot even register, we face a wavelike particle behavior moving at an extreme speed.

This feature (or what appeared to be a feature to our senses) of a particle may result from its almost infinitesimal size rather than its intrinsic value. To be more precise, a particle's inherent value is independent of our ability to perceive, measure, or comprehend it. We may grasp it to some extent or point, but its intrinsic value may still be beyond our ability and the ability of the technology or instruments we rely on.

Double Slit Experiment

When there is a detector, in the double-slit experiment, the fact that the detector is plugged in makes the waves collapse, but when it is unplugged, there is no interference with the detector, and the waves do not collapse. On the quantum level, we cannot undermine the impact of a plugged-in detector on a wave. Nothing happens when the detector is unplugged, and there is no impact on the waves. Once the detector is plugged in, it is a different story because it emits some of its “energy” (in one way or another), causing the waves to collapse and behave as particles. A plugged-in detector is in a different state when plugged in than when it is unplugged. On the quantum level, a plugged-in instrument's slightest changes or occurrences may affect the waves’ behavior in this experiment. A plugged-in instrument may cause a disturbance not just by pure immaterial recognition of a particle (or wave) but by its physical interference, regardless of how subtle or insignificant it is, with the waves causing them to collapse. This is not an observer’s effect but an effect of a possible disturbance caused by a plugged-in instrument.

WITHOUT MOTION, THERE IS NO GRAVITATION

MASS, MOTION, GRAVITATION

Matter is the product of a Universal Mind. Matter, as we perceive it, is nonexistent. The Universal Mind should not be equated with the fifth element, aether, as Aristotle or other ancient or modern philosophers and scientists understood it. However, it possesses some features of aether as they understood them.

We can almost be sure of two poles of the Absolute—Being and Nonbeing. Being, as I understand it, can be equated with the Universal Mind (Ultimate Mind) or God, provided we use the term God following this philosophy and not following its general use (as in religions), where this term serves the ideas, desires, and dogmas of the people who claimed to speak a word of God (and not to fit reality).

We can be sure that the fifth element (idea) was immaterial for Plato and Aristotle, who used the term aether. The fifth element (Latin: *quinta esentia*) differs from the other four elements (Earth, Water, Fire, and Air). When we look at aether, from the perspective of our philosophy, as the main principle before the formation of the world, as a potential (*in posse*), during its actualization (*in esse*), and as the underlying Being or reality of all the existence, then this term can be equated with God or, conditionally, with the Universal Mind. *A posse ad esse* is the transformation from the potential of the Universal Mind to its actualization as the Universe.

Although the Being (Universal Mind) is not material, it does not mean that we cannot, conditionally, call this Mind an immaterial "substance." This clarification is important to understand how an immaterial entity can transform into something we experience as material. Whatever we perceive and experience through our senses is based on conventions from secondary qualities of the world (as described by Locke, Berkeley, David Hume, and others). Perhaps our most admirable ability is primarily based on an "illusion." Without this illusion, the world would not only be a sad place but a place without purpose. The whole truth and the beauty of the world lie hidden in this illusion. Our Reality is an illusion, and we shall reinvestigate the word illusion. Without illusion, there is no reality. If illusion is the source of our reality, we shall redefine illusion.

Why not accept this illusion in all its glory and as it is: real that cannot be more real regardless of all the "tricks" we think it pulled on us? Without all the devices and "instruments" the Universal Mind or the Being "pulled off" to "create reality," we would not be able to experience all the beauty, glory, and miracles of existence. This illusion is the noblest thing reality can do for us. We must accept that we are real regardless of our awareness of what constitutes reality. Our knowledge does not make us less real. Our awareness shall not create more distance between us and the world. Quiet to the contrary, our awareness shall be a bridge for a fuller life and connection to the Ultimate source from which everything originates. This Source is not only the source of Everything but also the Source of peace and bliss. Even on an individual level, this realization leads to a better connection with oneself in light of the ultimate Oneness of Everything. Oneness is bliss in peace and peace in bliss.

This illusion is reality, and we shall acknowledge it as such. The interdependence of secondary and primary qualities, the dependence of our senses on the world, and the formation of our impressions are all realities. But, if reality is not reality, as we see it or understand it, this does not mean it is not a reality. Without these "illusions," there would be no meaningful reality. Reality as it is, in its ultimate and absolute state, without transformations, is equal to nothing.

The underlying truth of all existence and all secondary qualities of matter is the ever-present Universal Mind, Universal Primary Quality, which feeds all existence and creates and recreates matter itself.

We can accept energy transformation into mass and that they are the same. But matter, or this kind of energy, could have never come into Being just of itself and could not have created itself, as it is, from nothing. As we described, matter (energy, mass) is impossible without the primary quality. Not only would it not be possible, but it would also be dead without direction, purpose, and meaning. Although, according to Einstein, matter is a condensed energy, energy is still massless. Without kinetic energy, everything would not only come to a stop but disappear. Through motion, the Universal Source secures all the laws of physics, including gravitation and the universal cosmic order. In a way, energy is an unidentifiable "force," the Bridge between the universal Source and matter.

Richard Feynman[9] had to say this about energy in his 1961 lecture:

> "There is a fact, or if you wish, a *law*, governing all natural phenomena that are known to date. There is no known exception to this law – it is exact so far as we know. The law is called the *conservation of energy*. It states that there is a certain quantity, which we call energy that does not change in manifold changes which nature undergoes. That is a most abstract idea, because it is a mathematical principle; it says that there is a numerical quantity which does not change when something happens. It is not a

[9] Feynman, Richard (1918—1988) was an American theoretical physicist who received the Nobel Prize in Physics in 1965.

description of a mechanism, or anything concrete; it is just a strange fact that we can calculate some number and when we finish watching nature go through her tricks and calculate the number again, it is the same."

All significant philosophers and scientists throughout history were in their own right, right if we consider the context, time, and place, the point from which they observed the world by the means available to them. If we understand this context, we know how much harder it was for them to decipher the world previously unknown, except as an experience without fundamental and deeper understanding. In this sense, all these philosophers and scientists were, in a way, "right," even when they were "not" right. Correctness or wrongness of their ideas and opinions shall be measured more by how they helped our understanding and ideas develop directly from their thoughts. Even if they were in some way wrong, great ideas helped our ideas develop and allowed the formation and formulations of great ideas that will follow. Quality and potential of insights and ideas are more important than strict correctness without any potential.

Progress in human history would not be possible without following the traces of long-bygone giants (as Newton understood them). We can hardly produce any new important question that ancient Greek philosophers did not pose. The whole idea of Western philosophy, as it is, would not be possible without the ancient Greeks. This statement holds even when we talk about the modern era's greatest philosophers, starting with Descartes and culminating in the works of the great German philosophers Leibnitz, Kant, Hegel, Schopenhauer, the Dutch Spinoza, and others. Almost all central questions or problems treated by these philosophers were already postulated, discussed, or touched, directly or indirectly, by the great ancient philosophers who paved the way for the others.

Now, if we treat mass as a feature of matter, we will use these terms:

The total mass of the Universe = Tm
Infinitesimal mass = Im
Highest velocity (speed) = Hv
Equilibrium
Tm : Hv = Eqv (Equilibrium speed)
Tm : Im = Eqm (Equilibrium mass)
M × V = Tm (highest point)
Tm : AV = 0m (lowest point, no mass as a result of absolute speed)

MOTION IS ENERGY

At the lowest "absolute point" and the highest "absolute point," the speed becomes "absolute." At the "highest point," the hypothetical "end" of matter, the mass has no velocity (motion); it becomes static. The state of absolute speed is the static state. Absolute speed is the point of disappearance on the micro and macro levels.

The ultimate immaterial Source causes the first vibration, causing the motion, velocity, and gravitation of the "material" world that operates as the "program" of the Ultimate Source. This program does not mean that there is no free will. Without motion, there would be no gravitation, regardless of mass. Max Planck wrote: "All matter originates and exists only by virtue of a force which brings the particle of an atom to **vibration** and holds this most minute solar system of the atom together. We must assume behind this force the existence of a conscious and intelligent mind. This mind is the matrix of all matter."[10] He also said: "I regard consciousness as fundamental. I regard matter as derived from consciousness. **We cannot get behind consciousness**. Everything that we talk about, everything that we regard as existing, postulates consciousness."[11] We should combine this with what Erwin

---

[10] Planck, Max. "Religion and natural science," *Scientific Autobiography and Other Papers*, 187. The *Observer* (11 January 1931): *Psychic Research* (1931), Vol. 25, p. 91

[11] The *Observer* (25 January 1931)

Schrodinger had to say about the same subject: "Although I think that life may be the result of an accident, I do not think that of consciousness. Consciousness cannot be counted for in physical terms, for consciousness is absolutely fundamental. It cannot be accounted for in terms of anything else. Quantum physics reveals a basic oneness of the universe. Multiplicity is only apparent; in truth, there is only one mind."[12]

[12] The *Observer* (11 January 1931): *Psychic Research* (1931), Vol. 25, p. 91

UNIVERSAL MIND AND INFINITY

The road to existence and infinity as a potential (and to disappearance, for that matter) is Zero (passes through zero). Zero is the absolute passage from Nonbeing into the Being, essence into existence, existence into disappearance, and absolute finitude to infinity (infinite potential, infinite possibilities). (Existence is the Tower of Life, which is possible only between two zeros.)

Infinity does not exist in actuality. Infinity is only the potential of a Universal Mind. Since it is immaterial and not a slave to the space-time continuum and its properties, it is not dimensional and can be finite and infinite simultaneously. Since time and space have no meaning in the absolute state of a Universal Mind, then its finitude or infinity is almost irrelevant and of no importance.

Infinity is a mathematical, spatial, and temporal impossibility except as a concept. It is absurd if understood as an actuality (the universe, the world). Even if we try to imagine the infinity of the Universal Mind as "actuality" playing out all its potential simultaneously, that is impossible because infinity is both theoretically and practically unreachable.

The Universal Mind saves its uttermost purpose and meaning more through its potential than its actuality. Actuality can be one and many. All these real and potential actualities are limited, and their finitude will eventually make them disappear. Through its potential for infinity, the Universal Mind can exercise, not only simultaneously but perpetually, innumerable and always different (although based on the same laws) manifestations of becoming.

Although the Universal Mind is a Being, it is not concerned with becoming because it already is. Since the purpose and meaning of the Being, Universal Mind, as it is in its absolute state, is lost, it must either transform itself or produce from itself the world as we see, perceive, and experience it. This process, purpose, or true meaning may be called the rejuvenating process of the Ultimate Being by always becoming new through rebirths (in different ways following the potential) *ad infinitum.*

CHAPTER 5

WHAT IS THE UNIVERSE?

Is our idea about the Universe formed around what we think and know about the Universe, or is it based on what the Universe is as it is? We cannot know everything about the Universe either way. Still, we can sharpen our ideas and make them, to the highest possible degree, reflect the real Universe as it is and not only reflect our perception of the Universe. The more our ideas harmonize with what the Universe is, the more accurate this reflection will be.

From the standard view of the main religions, God created the Universe. Based on this standard, the Universe is material, but the Creator is immaterial. On the other hand, we can imagine that the Universe has always existed, and if that were the case, there was no creator to create it; "it simply is" (Bertrand Russell).

We certainly know that the Universe, regardless of whether it was created by God or not (always existing without a cause), is evolving. The Universe is not static. The Universe is the source, the cause, and an inexhaustible reservoir of energy, possibilities, and life. Although it sounds paradoxical, the Universe is "physical" and non-physical. As such, it contains metaphysics in its very Being. The physical feature of the Universe is only an expression of its metaphysical, "ethereal," nonphysical nature (the Kantian being-in-itself); physics is its appearance, and metaphysics is its essence. (The appearance is in motion, yet the essence is static. Motion [in the classical "physical" sense] is possible in the world of physics and impossible in metaphysics [immaterial world].)

Based on our perceptions and beliefs, the starting point cannot change the nature of the Universe. Created or uncreated, the Universe is. The Universe would never be different, regardless of our point of view; only our ideas about the Universe may change. The more important question is whether our concept of the Universe would be different if we changed our starting position. Could the Universe potentially be different depending on these two starting points? Either way, if God created it or it always existed in one form or another, the Universe may show and possess the same qualities, in which case this dichotomy would not be substantially important, except formally.

The third idea could imply God in the Universe (not in the strict sense of Spinoza's pantheism) and the Universe in God. What does this mean? It means that the Universe is, in either case, a manifestation of something that has always existed. If something never existed, it would not be able to come into Being. Absolute nothingness cannot give birth to anything, either God or the Universe. If this were the case, then Nothingness would be the first cause. If God is the first cause and source of everything, then based on this logic, God would be nothing because God came from nothing.

On the other hand, if the Universe came from nothing, the Universe would be nothing. Only nothing can come from nothing. Nothing is incapable of creating or making anything. Therefore, the question of who created God or who created the Universe is, at best, counterproductive and sterile.

From this hypothetical point of view, it would not matter if God created the Universe. If God or the Universe always existed in some way or another, the critical question would be whether there is any difference between God, understood in this way, and the Universe. For if God always existed, what would make it so distinctly and inherently different from the Universe? Or if the

Universe always existed, what would make it inherently different from God?

If we try to answer these questions, we will again be faced more with our inherited ideas about the Universe and God than with the reality and essence these terms should represent. Therefore, we should try to enrich and redefine these and many other terms we use. We, human beings, decided that the word (term, idea) God means, and should mean, something inherently different from the Universe. We also decided that the Universe means and should mean something different from God. But, if we, somewhat arbitrarily, determined not only the destiny of these terms and ideas but, based on them, our very conception and perception of what these ideas and terms are supposed to represent, one may ask how much closer these ideas and reasoning have brought us to the truth.

Once we have established our ideas and definitions, the main underlying question is whether our goal has been more in preserving and fighting for the preservation of our concepts and already established ideas as they are or in finding out if they represent the truth as it is and ought to be and not only as it is defined, arbitrarily declared, proclaimed or prescribed? If we get rid of all dogmas and established paradigms, we can conclude that what we seek must be the truth itself, regardless of how well or to what degree it would fit our views, concepts, and beliefs.

We must notice that religious thought, for instance, as presented by their most elaborate, learned, and sophisticated proponents, regardless of the grasp and superb philosophical knowledge (St. Augustine and St. Thomas Aquinas are extraordinary examples), did not, even with the best efforts to fight dogma and go from strictly philosophical positions, try to conceive God as something different from God as presented in Christianity or the *Bible*. A similar situation exists with the other predominant monotheistic religion, Islam. Based on this logic, as a subject of philosophical inquiry, God became *a priori* a question of how to establish the right to God more than how to set the truth (the right to the truth) itself.

The pursuit of truth (which must imply God, among other things) must lead to God if it exists. Nothing can be beyond the truth. If God is the ultimate principle, then this principle is the truth itself. The truth is independent and indifferent to the name or word we use to describe it, whether it be the Ultimate Source of Everything or God. If we look at God in this way, God equals the truth. But what spoils or may spoil our inquiry about God or truth, regardless of our sincerity, is not God or the truth itself but our distorted knowledge of both. (Also, the problem of language.)

The statement that either God is the ultimate cause of the Universe or that the Universe appeared from nothing contains falsity because there is a third modality, which is that the Universe may have created itself. But, if the Universe has created itself, it had existed before its creation. If it existed before, what would be the difference between this Being and God, or would it mean that this Being is something we traditionally have chosen to call God?

If there was Something, the Being, before the world's creation, the Being was uncreated. This uncreated Being is traditionally called God. The uncreated Being is the source of all creation, with the help of emptiness or absolute void. In that case, the Universe itself would be the Self-Creator, its own God.

WHAT POWERED THE BIG BANG?

Our discourse reveals the answer to what powered the Big Bang. If we claim that matter, as we see and understand it, is not matter, the conclusion is that matter (energy) is the product of an immaterial Being or Universal Mind.

We consider only the Universal Mind as the one possessing the primary quality, the essence of Being, which quality is not detectable by senses or indirectly by any means except by imagination and metaphysical speculation or thoughts. This immaterial Being, or "mind" (Universal Source), possesses the power to create through recreation and emanation of itself into existence. The requirement and essential elements of reality (as we see and perceive it) in the Universe are matter or energy.

Although energy, or matter, exists and we have already established our ideas about it, we still claim that matter does not exist except as a secondary quality of the universal mind. This transformation of the primary quality of the ultimate essence into the secondary quality of existence is the source of life that powers the Big Bang.

Nobody measured, and nobody will ever be able to precisely measure the "spot" from which the Big Bang started because it is a "spotless spot." Everyone knows that this spot is infinitesimally tiny. We claim that this spot is the zero spot (immeasurable) of the ultimate essence on its way into existence. This spot is immeasurable because it is zero. This "spot" is the spotless spot of absolute density and zero size.

Even if we naively try to explain the immeasurable spaces, energies, and galaxies of the Universe with the so-called dark matter and dark energy, it is hard to comprehend, regardless of all possible physical laws and laws of compression or contraction, that such waste energies can fit into a tiny "spot." If energy is indestructible, this beginning will prove the opposite based on its smallness. If something can disappear into nothing, it must be "destructible," regardless of our conceptions.

We do not claim that matter (energy) is either destructible or indestructible. We argue that matter is not real matter, as we see it, and therefore can transform into the primordial Being (Universal Mind) from which it came, or it can disappear if the "energy," power of its primary quality, still contained within matter, stops powering it.

Precisely in the same manner as our senses can taste, smell, and see, regardless of these qualities being nonexistent in matter itself, we can see and feel matter as something presented to our thoughts, feelings, and senses. We imagine that these other qualities, different from those described as secondary (related to the senses), are primary qualities of matter of the Universe. However, they are still only secondary qualities of a higher degree. Even if we categorize them differently, these qualities would still be secondary, and the qualities associated with the senses would be the tertiary qualities.

All we see is an "illusion" but a beautiful, meaningful, and purposeful illusion, no less real and realistic if it were the other way. The most important thing is the existence and rejuvenation of the ultimate, primordial immaterial essence. Transformation of this essence and power (supreme primary quality), similar from a human point of view to "alchemy," does not undermine the value and reality of matter and the Universe. In this sense, we believe that the ultimate immaterial Being is omnipotent and capable of creating and recreating itself through different modes of transformation. The most important transformation is the transformation of the essence into existence.

The ability to transform itself into different modes of existence is the richest source of the power of the Absolute. The Big Bang is initiated and powered by an Ultimate Mind that programs existence while offering freedom and free will from this source. This power only enables life but does not interfere with its development, which evolves into freedom.

Another force, which is not a force at all but is equally important, is the "force" of the Nonbeing, the force of the Nothing, the force of Emptiness, or the absolute vacuum. Without this “force,” space and time would not be possible. We must also state that time is impossible without space. In this sense, we treat time as a feature or property of space.

Space is only possible in existence because beyond existence, in the pure essence of the Absolute, there is no space, which means there is no time. The other mode or *sine qua non* of existence is a plurality. The world must transform from primordial oneness and singularity into the plurality of existence or life. This transformation does not mean that the singularity disappears but that this singularity transforms into a plurality that powers not only existence and life but, more importantly, meaning and purpose. Still, there is always an underlying Oneness pervading reality of which quantum entanglement is one of the most obvious signs.

Relationships are possible only in plurality. Only these relations can create distances, and only based on these distances from one another can space and time, in our sense, become possible. That is why there can be no space and time in singularity.

POINT OF CREATION OR RECREATION

The point of absolute equilibrium is where the Universe reaches its "highest absolute" point on the macro side and its "lowest absolute" point on the micro side. At this point, the macro and micro meet regardless of how paradoxical it sounds. The Universe is static at the absolute macro point, which must be zero. At the micro absolute point, which also must be zero, the Universe reaches its "absolute" speed—"omnipresence" secured by the smallest "immaterial particles." Absolute speed is static. In some way, this may be the basis of nonlocality in quantum mechanics. At this point, macro and micro unite. This is the point of creation and recreation, the source of infinity and finitude where the Being and Nonbeing meet to create the World. The Being is the active creative force, and the Nonbeing is the passive creative force. This "point" is the wormhole where existence and nonexistence meet to create the world, the

Tunnel from the absolute realm into the relative realm of the world and *vice versa*. Without one, there is no other. This point could also be the nucleus of a black hole at the point of its absolute density where everything sank into One without any space. That is the Zero point. This point is the gateway to new life. At this point, a black hole either "explodes" or disappears.

At the macro point, a similar thing happens due to the expansion of the Universe. Absolute expansion, precisely like a complete contraction, would lead to a similar result in the almost absolute speed, bringing everything to one "point"—the "point" of

Zero. This "point" is pointless; it is everywhere and nowhere. In this sense, this is the "point" where micro and macro meet and become the same. This "point" and "moment" is the source of the Big Bang. Precisely at the moment "where" and "when" it reaches "Zero," it explodes and recovers itself as a Big Bang. It is its new revival within the reservoir of absolute potential. (This does not necessarily mean that the end of one cosmic cycle or existence is the automatic beginning of a new one. On the eternal level, beyond time and existence in the material sense, every moment and point are the same regardless of how much time has passed or may pass in any potential material universe.)

If we would, hypothetically, explore the idea of the universe's contraction, the result would be similar to the black hole. Regardless of a much bigger mass of the whole Universe, as ours is, the absolute contraction would lead to the same point at which the black hole reaches maximum density. The point of maximum density is Zero, at which point the Big Bang happens (or the Universe disappears, which is less likely). In this sense, from the point of the result, expansion or contraction of the Universe would lead to almost the same result.

We believe it is more likely that there is not only one Big Bang but multiple Bing Bangs at different points in the Universe of Universes or Multiverse. Let's suppose there is more than one universe. In that case, we cannot talk about any of them independently of the rest, except hypothetically, because there would be no absolute void beyond one universe or Big Bang. In that sense, the Universe of Universes would be unaffected by anything beyond it except the Universal Mind or Absolute itself. All we state in this sense applies to the unknown or unknowable Universe exceeding our perceptive powers and scientific instruments.

Still, for argumentation, we can imagine there is only this Universe we are aware of, and we can treat it, along with black holes, as if there is nothing beyond it except absolute nothingness or absolute void. This world would be Einstein's space with the curvature of space. Beyond it, there is a "spaceless space" without curvature. But without the uncurved "spaceless space," curved space is impossible. This spaceless space is nothingness, and only a void provides room for the creation of space. The only real space is the spaceless space or absolute vacuum, and this spaceless space cannot be curved. What is curved is the creation that took "one part" of the primordial void, to put it that way, to create the World—the gravitational field and the spacetime continuum. But the primordial void, or the only real space that there can be, is not curved and cannot be curved.

CREATION

The most unscientific idea in the history of humankind is that something came (or can come) into existence from nothing. Yet, since the Absolute is nothing without its emanation (manifestation in plurality), it must create Something to be Something. Absolute is not nothing just for being nothing but because it is an absolute something at its "highest point," at which there is no need for further movement, and all meaning and purpose are lost. The moving of the same thing to a different same point is no movement at all and is pointless.

Any movement, even the movement of the Absolute, must have a purpose and meaning to justify it. From the point of the Absolute, it can only achieve purpose through creation. Creation is not any creation but implies the recreation of the Absolute through plurality.

The Absolute can only create and recreate itself if it transforms into something “less absolute.” This “less” absolute Being manifested through many beings is not only the Absolute's creation or recreation but is also its only life. In absolute Oneness, there is no real life. Division is the source of life. The secret to life is plurality. Oneness without plurality loses meaning. Plurality in Oneness is the source of meaning and purpose.

Life is not perfect. Almost nothing is ideal in the strictest sense, and we humans are not perfect in the strictest sense. But what is perfection? Even if a human being, or anything else, could have been perfect, would that "perfection" be desirable from an absolute point of view? Regardless of our debates about the intelligent design of the Universe or accidental appearance or always present matter and the always present Universe, the question remains—what is perfection (or absolute goodness), and how much do we know about perfection (is perfection perfect)?

We shall place everything in context. Purpose and meaning are the primary contexts or the basis from which we must start. Is the Universe without purpose and meaning? We shall try to understand the purpose and meaning better to answer this question. Then, the question would be—are purpose and meaning possible in any world? Could the world organize itself logically, meaningfully, and purposefully if it were an utterly dead matter? Is it possible that, by pure chance, only based on some accidental possibilities, the Universe would be able to organize itself, regardless of its size and the number of its potential units or micro-universes?

It seems reasonable to believe that, regardless of the number of possibilities and universes, the Universe (Omniverse, Macro-Universe, Multiverse) would still be nothing more than a dead thing. What is it that powers the Universe? We see and recognize motion; we recognize the four main forces (gravity, electromagnetism, and the strong and weak nuclear force) of the Universe, but we still do not know or fully understand these forces or the Universe itself. We recognize the Universe; we can feel it, touch it, think about it; we can describe it and its forces, but do we understand it? If we admit that we do not understand it fully, we must acknowledge that we do not fully understand perfection. Suppose we realize that we do not fully understand perfection; we must shift the subject of discussion to a higher level, which is not necessarily a higher level of understanding but a higher level regarding our approach.

Suppose our primary goal is to win the argument, attract attention, and popularize our ideas at any cost, even though our efforts may still be noble. In this case, we cannot completely count on a higher level of understanding. Then, the question of nobility arises. Someone may possess the noblest of spirits but still not be capable of the best results based on endeavors and thoughts. The other question is, are the best thoughts the noblest thoughts? Or can the best thoughts be produced by the spirits not so noble? On the superficial level, all these unimportant ideas may lead to the highest level of understanding through constant questioning of everything and everybody, pursuing the "ultimate understanding."

We may win and still be wrong or partially wrong. We may win based on the merits of a superior thought at any given moment. But would we have won if there was a thought superior enough to challenge the winning thought at any particular moment? We would fail if there were a superior thought at any given time to challenge us. Would even the more superior idea be nobler or only superior in terms of the truth? Countless questions always arise and follow any serious inquiry.

How do we know our thoughts are the best and, simultaneously, the noblest? How rigorous is the inspection of our thoughts and merits? Are we satisfied more with winning or knowing? Do we enjoy victories and fame more than the truth itself? Our motives are driven more by winning than by finding the truth. If when this is the case, the winning "truth" or personal "triumph" can be more devastating, not only to “progress” but also to knowledge and the truth, than if there were no such "truths" and such thoughts at a particular moment. Some ideas may temporarily open some doors but close others in the long run. (Features of this kind in human nature are most apparent in politics.)

We are often “enslaved” more, not by the merits of ideas or the level of truth they contain, but by a paradigm shift, habits, and even the way of living that some ideas may have the power to create. We can easily accommodate even our language to the prevailing ideas, our way of thinking, our habitual modes, our systems of learning (not only understanding), and our system of thought.

Considering all these elements, the context and frame from which we try to think or place ideas become more extensive and cannot be easily simplified to fit the purpose of a "winning" argument or idea. For example, the concept of intelligent design is, in a strict sense, a religious concept, but must it be strictly religious? To answer this, we must first ask what intelligent design is. And then, what may this design be from a metaphysical point of view, from the point of view of creation or recreation? How much do our particular human ideas about design enslave us, and can we even think about the design outside of our strictly human context and place it in an extraordinary context of the Absolute, God, existence, essence, matter, and Universe?

If we do this, we may realize that we often apply our ordinary words, as the design is, to extraordinary contexts without any modifications in thinking or linguistically, not to mention understanding. For that reason, if we want to think about the intelligent design concept (concerning God), regardless of our purpose, whether to prove or disprove the existence of God, we cannot do this in a serious or academically acceptable manner, without redefining the words design and God, without enriching them with some new meanings to fit the unique and extraordinary context.

Finally, if we do all this, we may come to some extraordinary or even "absurd" conclusions on the superficial level. For instance, one of these conclusions about the intelligent design concept is that Deism and Atheism, Belief and Disbelief, are not inherently incompatible as they seem at first sight and on a superficial level.

DARK MATTER

The visible universe contains only 5% of visible matter, whereas dark matter constitutes 27%, and 68% is dark energy. According to some scientists, dark matter may be in other dimensions. This idea is compatible with my idea of an Omniverse, although a Multiverse does not need to be in other dimensions. Nevertheless, such a Multiverse may contain material (perhaps unapproachable by human beings) beyond our imagination.

Since scientists have no answer to dark matter (except dark matter filament), the most logical explanation is that it is invisible. Some questions remain:

1. Can matter be invisible (or imperceptible by our senses and instruments)?
2. Even if matter could be theoretically invisible, is it possible that such a vast amount of matter, like dark matter, would escape all our knowledge and existing laws of physics and be unidentified until recently but wholly invisible and beyond our reach?
3. If dark matter is imperceptible, what makes it imperceptible?
4. Is it potentially perceptible but not perceptible to us as human beings?
5. Is there anything that would still avoid perception even if we possessed the absolute perceptive ability or technology with these abilities?
6. Or, is dark matter our way of explaining the unexplainable and offering a linguistic form to unknown phenomena?

7. Or, is it our inability to go beyond the spectrum, outside the existing frames, and try to decipher the unknown beyond the known frame of reality or what we see and understand as reality and the Universe?

The answer to the first question is known; even atoms are invisible not only to the eyes but to microscopes. It is, therefore, theoretically possible that matter can be hidden and imperceptible. Still, it is hard to imagine that vast amounts of the mass of the Universe would stay unaccounted for within the realm of already advanced understanding of the laws of physics, instruments, and experiments. It would be possible to prove mathematically, based on what we already know about the Universe, the mass, the dispersion of energy and mass, and by these comparisons to conclude, without the CERN accelerator, that this is, most likely, impossible. This was a short answer to the second question.

The third question is important because it would lead scientists in the right direction by avoiding the possible net of perplexed ideas. If we have already established that something exists, it would be better to define it as precisely as possible to avoid guessing only. In addition, how do we guess? We do not know anything about its nature, origin, or how it came into existence except that we came to this discovery almost accidentally by pure and relatively simple measurements and experiments. But what about us? How do we think? What methods do we use in experiments and the way we think? The answer to these questions could lead to better discoveries than only focusing on something we do not know and, even worse, if we do not know where to look for it.

Based on an accidental discovery, it is a good start to conclude that there is more mass in the Universe than can be detected. Still, it would be better and more productive to go beyond the Universe as we see it, beyond our existing knowledge and perception, not toward the stars we already know but toward another bottomless sky of darkness and the unknown. Although light is the source of life, darkness is also the source of light and life. Maybe the brightest "star" sleeps in the darkness and feeds the world from darkness.

Is there only one Universe?

If we start from the premise of the Big Bang theory, it would be logical to ask why there is only one Big Bang. It is easy to conclude that if there is a Big Bang at one point in "space" (nothingness), there can be another one at another "point," past or future, although this may sound strange.

DARK ENERGY

The source of 68% of the mass-energy density is dark energy.

The Universe is one, even if we use the word Multiverse. If it is not one, then it is not a Universe. If what we understand as the Universe is not one, it does not mean that there are multiple universes but that our ideas about the Universe may need to be corrected or that they do not quite correctly represent reality. If this is the case, we should adapt the language to reality, not the reality to language.

If we accept a Big Bang (although our view would be the same regardless of this premise), we must accept the possibility of multiple Big Bangs anywhere else. We can imagine many more (some even bigger) Big Bangs and "universes" (like our own) than there are galaxies in our Universe. Now, we will name this Universe, containing all others, Omniverse (Omni-Universe) or Macro Universe. Dark energy is the "unmeasurable" energy of the Omniverse (Macro Universe).

If we imagined "space" before any universe came into Being, there would be only an absolute vacuum. Absolute vacuum is not "contaminated" by anything; no matter, no energy, everything is pure because only nothing can be completely pure. Primordial vacuum, as we know, is not space because where every imaginary point is the same, there is no point and no distance from anything to anything; this means there is only a zero.

But without this zero, no point would be possible, and no distance from anything to anything would appear since there would be no points of reference (no objects). Absolute vacuum is potential space and potential for space. Only when something and nothing (absolute vacuum) interact do something and space show their real nature and life.

If this something is only the Universe we are able and capable of experiencing, or if we imagine the Universe, which is, most likely, the real, much bigger one, the space would still bear the characteristics and "colors" of the one we can experience. Beyond this realm, there would be no other point of reference (Universe), and this Universe would be only a speck in the realm of nothingness. Even if this Universe emanates some of its fundamental forces into the realm of nothingness, this would still be of minimal reach regarding the potential "vastness" of "space" (vacuum) outside that universe. With one Universe like this, there would still be no distance between it and anything else and outside of it (or, precisely, after some point outside of it); time and space, as we understand them, would not exist.

Regardless of how far or how close any hypothetical point in space and time would be, it would still be the same point because nothingness is the same everywhere. The only point of nothingness is nothingness itself. There is no real space beyond nothingness and no real-time. The only way for nothingness to "survive" or "outlive" its nothingness is to live with the Being and move into the Being. The symbiosis of the Being with the Nonbeing is the source of life and space and time (space-time continuum).

If there is only one small Universe, as ours is, in comparison to the vast potential of the Being and the Nonbeing, then this Universe would be smaller, in comparison to this potential, than quark is in comparison to our Universe. At this point, we will bring in Einstein's Cosmological Constant ("biggest blunder") to clarify and present this idea within the context of dark energy and the Macro-Universe.

Suppose we use cosmological constant interchangeably with dark energy. That leads us to conclude that this kind of a universe, like our own if limited to our perception of it, would not be a sound basis for explaining dark energy or cosmological constant within the realm of the Absolute or its potential. The real meaning of the Absolute is not the Absolute itself but its potential. Its potential is both in the Being and in the Nonbeing.

Not only the "infinitesimal" Universe like ours (from the point of the Absolute) would not meet the standard of the cosmological constant, it would not explain the dark energy, but it also would not meet the standard of nothingness or the absolute vacuum concerning their potential respectively. Our Universe does not even remotely reach the unlimited potential as one manifestation of the Absolute.

There are two absolute potentials of the Absolute: Absolute potential for particular manifestations and absolute potential for infinity. No specific thing can be infinite, and no particular thing can be absolute. Nevertheless, anything can possess absolute value within and serve the system. The absolute potential is not in its infinity but in its potential for infinity. Within the realm of the Absolute, if we exclude the material universe, there is no space and time, and our concepts of infinity or finitude are mostly related to our view of space-time relations.

If there is only one Universe, like ours, it is too small and does not allow for exercising the highest possible, absolute potential at a given moment. For the highest possible absolute potential, viewed through infinity, to be exercised, there must be not only an infinity of possible worlds (universes) at different points (times) but the highest possible number (in this case, quantity is the main quality) of worlds at the same time at any time. Suppose there is no highest possible number of worlds (not only in variations but structurally). Under such conditions, the highest possible potential at any given time would be impossible because the chance, the main source of a potential infinity, would be unable to function and exercise itself to the highest possible potential.

Chance is the source of potential infinity and the primary source of real meaning in the manifestation (Universe) of the Absolute. Regardless of the unlimited potential for variations in quantity and quality within one universe as we understand it, such a world would still be limited in its manifestation and meaning if it were not a part of the Omniverse.

In a physical sense, the basis and start of existence and life is when the Nonbeing "moves" into and inhabits the Being (Being envelopes the Nonbeing). When Being and Nonbeing are divorced, there is no existence or life as we understand it.

Dark energy is the "unidentifiable" energy of the Omniverse. Dark energy comes from space outside *the observable universe or a sphere with a diameter of about 28.5 gigaparsecs (93 billion light-years or* 8.8×10^{26} *m).*

FINE-TUNING

For the fine-tuning to be accidental, there would have to be a multiverse and an "infinite" number of universes. Nobody has yet measured the level of fine-tuning in objective and absolute mathematical terms. This idea would imply calculating everything on every level, micro and macro, in every direction and combining everything with everything. Most likely, this would show that only the "infinite" number of universes would provide the real source for such a possibility because the level of fine-tuning in the existing Universe is so fine that it would require almost an infinite number of combinations to organize one in such a manner. At the same time, similar fine-tuning can happen in some other universe. The vast potential goes far beyond the human scope of comprehension, understanding, and observation. However, infinity as such is impossible mathematically or logically. Infinity only exists as a potential.

If we calculated the possibility of a googolplex[13] possible accidental universes (created somehow just of themselves) as ours is and divided it by the needed level of fine-tuning, we would find out that any imaginable number would not be enough to organize and sustain itself to the level of fine-tuning required for the Universe as ours is or any similar universe. In other words, its complexity, fine-tuning, or structure almost borders infinity. Any

[13] $1 \times 10^{10^{100}}$ (It would take more than the age of the Universe to write it down.).

imaginable number of universes would not be enough to account for all possible variations needed for only one finely-tuned Universe.

Perhaps it would be easy to calculate that the possible number of universes would always be less than the number of variations needed for one finely-tuned Universe to take place in such form. This number would have to be infinite, and that is impossible.

Nevertheless, how is this universe possible if infinite variations are needed? The whole meaning and beauty lie in the fact that it is possible and finely tuned. However, it can only be new in a new birth or rebirth to secure an infinite development and meaning of existence and life through chance—the source of infinity or endless potential. Otherwise, it would always be the same or a wholly programmed different universe without free will, but our Universe has free will. Chance itself is the source of infinity and the potential for variations.

- The absolute "number" of possibilities is the Absolute itself because it contains this potential.
- The absolute "number" of possibilities is infinity itself.
- The absolute "number" of possibilities is needed not for infinity but for any particular manifestation of the Absolute in the form such as the Universe.
- Any such universe, or the manifestation of the Absolute, requires many possibilities to exist meaningfully as a high complexity.
- The absolute number of possibilities is absolute potential.
- The potential of the Absolute is both infinite and eternal.
- The absolute "number" is infinity.
- Absolute "number" is numberless.
- Infinity is nonexistent. It is zero.
- Zero is a gateway.
- Zero is the Wormhole from the Universal Mind to the "Material" World-Universe.

- Possibilities are possible only when they are not zero.
- Passage through the Zero is the birth of possibilities.
- The present is an eternity.
- The victory of the finite possibility over infinity is the birth of life and existence.
- Victory over eternity (absolute time or space) is time's birth.
- Victory over the infinite space of zero is the birth of space.
- The Finitude of the Being makes infinity.
- Infinity in itself is nothing.
- Infinity of the Being is a never-ending process, life or existence.

Energy or matter, without being powered by the Ultimate Universal Force (Universal Mind), the source of everything, would not be able to finely tune itself irrespective of the potential for combinations and irrespective of its size (even if it were millions of times larger than our Universe). If it were not for the force beyond energy and matter (the Ultimate Immaterial Source of Everything) feeding, organizing, and fine-tuning it, energy and matter would be dead, without any life or force in them.

UNITY

There is Something, the Being, and this Being cannot be denied empirically or theoretically and cannot be challenged by skepticism. It would also be hard to deny that the Being (Universe) may be God himself (Itself), theoretically or hypothetically. If, for a moment, we forget the idea of intelligent design or the idea of a benevolent God or any God as presented by religions, and even if we forget the idea of a God who creates the world (Creator-God), we can still think of a God who recreates himself in the form of the Universe.

The idea of intelligent design or fine-tuning of the Universe and its high complexity must be reevaluated more in the sense of a God-Universe than in understanding its complexity and probability. First, if the Universe exists, there is proof not only of its probability but also of its existence. It does not matter how finely tuned the Universe is; what matters is—*how* this is possible and not *if* it is probable. We already know that it is probable, and that is enough. To speculate about the possibility of probability on a universal level (micro and macro, metaphysical and physical) is a waste of time in a race for the sake of the race to find a winning argument and not the winning truth.

What if God is here? What if God is the Universe? If we ask this question, it becomes clear that God is the subject and must be the subject of scientific inquiry equally, if not more, as of religious investigations, not to mention philosophical ones. Arguments about the existence or nonexistence of God should not be based only on assumptions and premises that God is something beyond the World, beyond the Universe. It is equally legitimate to consider the Universe, God-World, as the God himself (itself).

It is impossible that any (incredibly highly complex) universe would be, or can be, created by pure chance. If any chance is involved, then this is a chance of higher order and functioning under the ultimate laws of the Being. But this chance (as we understand the word chance and use it) is zero because "chance" gives a chance to the possibility (probability). The chance is the creation itself (the moment of creation) and is not random. Chance is responsible more for improbability than for securing probability because the driving force ("engine") of existence is not a chance; it is not evolution *per se* but the potential activated through evolution and not caused or created by evolution. Evolution manifests degrees of existing potential (Being, Absolute Mind). (This potential is the infinity of probabilities [which excludes improbability because if there were improbability, there would be no infinity].) We are all part of the paradoxical labyrinth (infinity) of the strange, mysterious being called the Absolute. Solving this biggest mystery of all helps us solve our own mystery of existence because the Absolute is one organism of which we are minuscule cells.

It is impossible that, by pure chance, the Universe would organize itself in any way (not to mention finely tuned) either in one Universe or a multiverse. Per a cosmological constant, as we see it, the Universe must contain the same potential everywhere, and this potential is not dependent on chance. Still, it is enriched by chance so that the infinite potential may work along with free will.

Without this potential (eternal, ever-existing mind), a chance would have no chance. The whole world would only be a matter of dead matter and an accumulation of accidental events with a zero chance of ever becoming anything remotely organized as the Universe is. This scenario would be equal to nothingness.

As we previously postulated, God of itself is nothing, and the Nothing of itself is nothing. In the absolute, without the world's existence as a plurality, God equals nothingness, and therefore the whole Absolute is nothing. The world is the Exit of the Absolute through God into nothingness and, in that way, into existence. The primordial essence in this manner becomes existence. Once the Being (God) and the Nonbeing (nothing) are united, neither one is nothing anymore. In this way, not only does the Absolute save itself, but it also saves the Being and the Nonbeing. This Nonbeing (nothing) is not nothing anymore but becomes something equally crucial as the Being itself.

The so-called "organized complexity" should not be ascribed to God or the Universe but to the complexity of our understanding. These complexities are not complex *per se* but arise from our limited powers of comprehension and understanding. Not only is God (the Being) simple, but the Universe is also simple. Our idea of complexity and the complexity of our knowledge is not simple. Whatever we do not understand becomes complex but becomes simple once we understand it. The level of our understanding is not the measure of complexities but our abilities. Complexities are proportionate to our abilities: the bigger the abilities, the lesser the complexities.

Almost perfect precision and fine-tuning, containing laws and order on every level in the Universe, proves more simplicity than complexity. Simplicity is the safer way to order than anarchy, which is complex precisely for its lack of order. What is complex to us is not complex to God. Not only is God simple, but the world (Universe) is also simple in his mind (the world's mind at the same time), which is ours too. We may also say that God and the world are complex from our point of view and simple from God's point of view (which does not help much if we only try to argue for the sake of an argument and not for understanding).

We cannot apply our complexity criteria to God because everything must be inherently effortless and simple for God;

otherwise, there would be no God. If there is a God, he cannot function based on the chaotic idea of trials and errors and leave everything to chance. The role of chance is limited to exercising the universal potential for infinity and free will.

The question, "Is God a necessary being?" can be reformulated to read: Is the Being a necessary being? If there is a Being and we know there is (the world), then it must be necessary; otherwise, it would not exist. Not only is the Being necessary, but it is also indestructible and eternal. If the Absolute Being were only necessary and not indestructible and eternal, it would not exist because necessity would not create it. If there is nothing, there is only nothing. In nothingness, there is no necessity out of which anything may come into the Being. If the Being is necessary, this necessity is always present as the Being itself.

The necessity of nothingness does not create anything. The real value and meaning of the Being are not in its necessity but in its very existence. If it did not exist, necessity would neither help nor create it. If the Being exists, and it does, it is pointless and redundant to insist on its necessity. The necessity of itself, without the Being, is nothing. The main point is if there is a Being, not if it is necessary. (What is) obvious need not be proved. Explanation of the Being and the creation or the existence of the world is not in its necessity but its existence. The Being is unborn and always existed regardless of being necessary or unnecessary.

The Unified Field (Theory) of Everything

A force feeding all the "energy and matter" in the Universe, creating the unified field, is the Universal Mind (Universal Source). Michio Kaku said that "an equation an inch long would allow us to read the mind of God." Clerk Maxwell created the first field theory (electromagnetism) in the mid-19th century. In the 20th century, Einstein's general relativity theory, treating gravitation, was the second field theory. Einstein and some other scientists tried to create a theory where electromagnetism and gravity would be different aspects of the same field, but they failed. With quantum physics, things became even more complicated. Einstein tried to reconcile the theory of relativity and quantum theory with no success.

The standard model theory describes electromagnetism and strong and weak nuclear force, yet gravitation stays out of the picture. I must reiterate that gravitation is not a force and cannot fit any equation. The "force" affecting all matter and energy, as described by the law of gravitation, is not the consequence of mass but of motion. The "force" of motion of larger masses is larger than the "force" of motion of smaller masses. The mass does not cause gravitation and is not the cause of attraction and curvature of space but motion. The Universal Mind in action is the Ultimate Unifying Force, the Universal Source of all information needed to create energy and matter, the laws of physics, and the Unified Field of Everything. The Primordial Immaterial Indestructible Energy is the Universal Source *powering* the Unified Field of Everything.

Oneness: Unity between the Universal Mind and the World

The divided One is still the One. The undivided One is the One that sleeps. The *modus operandi* of the One is union in division and division in the union. There is no escape from the One. No division can damage the oneness or wholeness of the One. There is only one Organism. Although the Source is in the sphere of meta or transcendence, it is still the One with the World. Creation is not separate from the Creator. It is the Creator itself. Not only can the World be separated from the Creator and stay One simultaneously, but the World creates and powers the Creator in the opposite direction. The Source of Everything powers Everything, and Everything powers the Source. Without the Source, there is no World; without the World, there is no Source. The Purpose is the

Source of the Source. The purpose is in the division and creation. The life of the Source is the World.

STATIC UNIVERSE AND ONENESS

“Absolute” velocity happens when the primary quality of all existence or Universal Mind shows its power in action, being everywhere simultaneously, not only faster than the speed of light but at the absolute speed, which is omnipresence. On this level, the time is absolute, and there is no relativity. Relativity of time is possible within the "visible" realm of reality and not in the manifestation (action) of the primary quality of a Universal Mind. Primary quality functions within the realm of Zero, securing omnipresence in the eternal present. Eternity is an absolute present. Relativity of time is possible only if there are the past and the future, not the present. The present is frozen. The present is Zero, and Zero is absolute in its way.

On the level below the strings, the smallest immaterial “particles” (“waves”) reach the "absolute speed." Based on this, the smallest and undetectable “particles” (“waves”) are everywhere, micro and macro, since they are fundamental and indivisible. They must be anywhere simultaneously at an almost absolute speed. Absolute speed is equal to no speed. That is the static state of absolute speed. In this way, the Universe is “static” on its “highest” “point” on a macro level and its lowest on a micro level. From this principle comes the unity of Everything. At the moment of “absolute speed,” energy or matter disappears.

The Original Universal Force, the Creator of Everything, unites the whole Universe in a single organism (the same in every direction and at any time). On the immaterial level of Reality,

Einstein's "Cosmological Constant" [Λ] may function as Absolute Oneness. The underlying Reality encompassing Everything is an invisible singularity (oneness).

Quantum entanglement

Quantum entanglement[14] is a sign and subtle proof of the Universal Mind securing the oneness or singularity of the "matcrial" world. This way, the ultimate immaterial reality operates within the "material" reality. The laws of physics are at work with maximum speed, yet the information is ever-present and omnipresent simultaneously, which is, in a way, proof of quantum entanglement.

Einstein described the phenomenon of quantum entanglement (particle entanglement, when a particle is in two places simultaneously) as the "spooky action at a distance." He also said: "I don't believe in quantum physics because I believe the Moon is there even if I am not looking at it." We agree with Einstein that the Moon is there, whether we look at it or not. We also believe that we cannot affect the position of a particle, either if we look at it or not. It appears as a particle when we look at it because we identify and recognize it. When we don't look at it, it is a wave, an illusion. We do not make any impact on it. We also agree with Einstein that a particle has a definite spin before being measured.

Einstein has shown that one particle can affect the other if the signal travels between them faster than the speed of light.[15] On the macro level, if we disregard the micro level of a micro level at which the smallest immaterial indivisible "particles" (not yet discovered) may travel faster than the speed of light, the signal traveling between the "particles" may be faster than the speed of light. It may be at any place at any time. Information is not lost

[14] Erwin Schrödinger conceived the term quantum entanglement in 1932.
[15] This is Universal Mind in action. Every place becomes the same place. Absolute speed.

because, on a micro-level of a micro-level, there is an "absolute" velocity (of immaterial “substance,” “particles,” information, messages, “thoughts,” and underlying oneness of reality) that secures the “absolute spacetime” even in the world of plurality which the Universe is. This principle is oneness in a plurality (or singularity in diversity). In this way, the original oneness (singularity) of the primordial Universal Mind of the Absolute is saved even in the world as its manifestation.

There can be no plurality without oneness (singularity) simultaneously; otherwise, the world would not be possible. Without the underlying oneness, the world would be a mechanical compilation of "dead matter," incapable of producing any logical or sustainable physical system or the world, not to mention biology and life. Without oneness or singularity, the world would be “existence” without existing, equal to nothingness.

Quantum entanglement, securing the instantaneous interconnectivity among the unimaginable number of “material” entities, is the underlying force in action, uniting everything in one superbly interconnected and alive organism. Quantum entanglement also manifests “absolute speed” and nonlocality; otherwise, particles would not have instantaneous interconnectivity. The Universal Mind (Primordial Immaterial Force) is the uniting force of everything. In partnership with emptiness, the Universal Mind is the creator of everything and reality as we see it.

- $A = \infty p$ (Where p is potential) Absolute is infinite potential.[16]
  $0 = \infty$

[16] Although I predominantly use my own symbols, mathematicians can modify them in the future, along with formulae, if needed. I am aware, for example, that p is momentum, not potential, but with some liberty, we used it to represent potential in our case. Likewise, we used other symbols in different situations.

- W = P (Where W is the world or U—universe and P is plurality)
- A = P+p (Plurality)

Any existing world (Universe) is finite. The Absolute Mind is immaterial and limitless, but it is still limited in itself and any particular manifestation (the world) and infinite in its potential. In other words, it can appear at any time, anywhere, as a specific manifestation, and it can go on (appearing and disappearing in the form of universes) forever *ad infinitum*. This potential of the Absolute (and the Nothing) for infinity (as a never-achievable goal) is the leading cause (source) of uncertainty, which, ultimately, is the source and basis of free will. Without uncertainty, there is no free will. Determinism excludes uncertainty and, therefore, free will.

On the other hand, the world is determined by its essence (the Universal Mind's "immaterial substance") and by its universal laws that are the same everywhere and are determined. If they are determined, that implies no uncertainty, but this conclusion is incorrect. Regardless of the essence from which everything comes to be (or is transformed), its manifestation, or its material substance, the world (Universe) will constantly evolve differently in any particular manifestation. The starting point, if nothing else, will always be different and form different universes in every specific manifestation.

Both the Being and Nonbeing are limitless, which makes them "infinite," and infinity is zero. Their frame is the Absolute itself, simultaneously providing the absolute limit and potential (infinity). The purpose of the world and the Absolute is not one particular existence and one particular world (Universe) or one life but a continuation of existence (universes) *ad infinitum*. Without infinity, there would be no real life because, at one point, existence would cease to exist forever, and it would be as if it had never existed. But infinity (although never reachable) secures continuation. Any manifestation of the Absolute, in the form of the World, follows the uncertainty principle, preserving the youth of the Absolute and originality (uniqueness) of any or all universes existing simultaneously or at any past or future time.

"Dimensions" that we can never explore are not some exotic dimensions of the string theory but more the "dimensions" of the micro-micro-level and macro-macro macro level that humans will probably never reach.

We can bypass the laws of physics only if we bypass the consequences of physical transformation. If we could operate on the level of pure information and return to the material form, not only would time travel be possible, but more valuable and fundamental endeavors would also be possible. When we say time travel, we do not mean traveling to the past or future using a time machine. That would be something similar to alchemy. Perhaps humankind, or any other species, can never achieve this level of science and technology except God. But we can imagine Beings transcending their physical existence and becoming immaterial. We can also imagine such beings on other levels beyond the "physical" world and existence known to us. The Ultimate Source, the Universal Mind, is the Ultimate Alchemist of Reality and the Universe.

CHAPTER 6

COGITO ERGO SUM ARGUMENT

Parmenides said: "To think and to be are one and the same." Philosophers had similar thoughts about this question from Plato, Aristotle, and Saint Augustine to Avicenna. "*Je pense, donc je suis*" ("I think, therefore I am, "or "I am thinking, therefore I exist.") Descartes used first in French in his *Discourse on the Method* (1637) and later in Latin in *Principles of Philosophy* and *Meditations on First Philosophy*.

One of the easiest ways to reconcile Descartes' *cogito ergo sum*[17] argument with counterarguments against it would be to modify it slightly:

> I am the thought.
> This thought exists.
> Therefore, I exist.

Everything that exists, regardless of whether it is aware of its existence, is information itself, a message, or a thought of the Universal Eternal Source of everything. The I that thinks, whatever it may be, exists. An I is not the source of thinking, but thinking is the source of an I. An I is the consequence of thinking. An I does not presuppose existence but is only a confirmation of existence.

---

[17] In *The Search for Truth by Natural Light*, published posthumously, Descartes formulated his idea as *dubito, ergo sum, vel, quod idem est, cogito, ergo sum* ("I doubt, therefore I am—or what is the same—I think, therefore I am"). This unfinished work (philosophical dialogue) was written in French but was first published in Duch translation in Amsterdam (1684).

Existence is not the consequence of an I. I do not exist because I am an I. Thinking I is a confirmation of existence *per se,* independent of whether I am that I or not. The sole possibility that I may think I am thinking is enough to prove the existence of a being that thinks or thinks that it (he-she) thinks. Otherwise, this being would not be able to be wrong or right, delusional, or deceived. Identification of an I, and with an I, or with the self, is not the source of existence: I do not exist because I think, but my thinking, even if not mine, proves the existence of whatever or whoever is thinking.

Thinking is not bulletproof, encompassing all of existence because existence contains many more things than there are thinking things. Nevertheless, nothing exists that is not a result (product) of a Universal Mind. The underlying reality of everything is the "Mind," irrespective of whether the majority of things making up the reality are aware.

Even if an I is an illusion, we cannot deny the existence of that illusion. Reality is an "illusion," which is not less tangible for being an illusion. If it were not an illusion, it would not have been possible as this kind of "reality." Different, or substantially different, reality would lose its purpose. The purpose of reality is not perfection or reality *per* se but the voyage toward perfection and reality with meaning.

We have to ensure we understand what existence is and what an I is in the context of an accepted consensus about these terms and definitions. Our reasoning and arguments may be correct if we know more deeply what existence or an I is. The most important thing is to go beyond words or literal expressions to catch the real intentions of philosophers and thinkers rather than to catch potential linguistic errors. We may temporarily win arguments and make personal gains if we only pursue linguistic errors. Still, we would produce confusion and lead the sincere search for scientific or philosophical discoveries astray.

The majority of contra arguments, from Pierre Gassendi, Georg Lichtenberg, Søren Kierkegaard, Friedrich Nietzsche, Martin Heidegger, John Macmurray to Bernard Williams, come from the idea that Descartes presupposes an I doing the thinking, although this may not be true. (Another significant counterargument, relating to the Decartes' method in general, is that he never questioned the doubt itself.)

What is existence? Existence is any state of the Being. Matter as it is, unaware of itself, exists regardless of "not" knowing that it exists. Still, as a part of a larger whole, any particle of matter contains information that serves that particle's specific purpose and the whole's purpose. Only nothing is not existence. But, without Nothing, existence would not be possible, so the Nothing is an essential part of existence. Still, we may say that only existing with some awareness is worth living.

We believe that Descartes was more interested in proving existence *per se* than his existence based on his identity or thought of his identity. He was interested in existence and thought *per se,* and an I is an accidental consequence of something that exists. I could be anything and could be an illusion. That is not the point. The point is that this I, regardless of how delusional or even if it were an illusion, is still something that can think He thinks, proving that "He" is, regardless of whether he is an illusion. Even an illusion is an existence. To be an illusion is to be, too.

We cannot be sure we know what reality is. We must find out what reality is. The majority of what we see is an illusion. The foundation of reality is our illusions about it, regardless of how paradoxical this sounds. Reality is an illusion, which does not make it any less real or valuable. In this reality, there is no Cartesian dualism since the underlying reality of everything is a Universal Mind. Matter is just a construct of the Universal Mind and is a valuable illusion; it is more useful and realistic if we know this.

For existence, it is not necessary that something must be "real" in our sense of the word but that it exists. Anything that exists, be it an "illusion," is existence. Anything that can think about this existence, and this "reality" or "illusion," can identify with it, which confirms its existence regardless of how distorted it is—the existence itself, the thinking, and then I thinking the thinking. That "I," whatever it may be, which is doing the thinking, even if it is "not" Descartes, exists. That is the whole point. It does not matter who is doing the thinking. What matters is that the being capable of recognizing this thinking, irrespective of who is doing the thinking, confirms its "own" (whatever it may be) existence; otherwise, it would not be able to be wrong, deceived, or anything else.

All that thinks or believes it thinks exists.
I think I am an I and exist even if I am not an I.
Existence is independent of personality.

Not everything that exists thinks.
Nonthinking does not necessarily equate to nonexisting.
But all that exists is powered by the Universal Mind.

We can solve this problem by identifying thought with existence based on our idea that everything is a "thought" (information) and part of the Universal Mind. Even if my thought, strictly speaking, is not mine—if "I" am the thought or information, "I" at least exists as a thought or information (regardless of who or what an I is).

But what about thinking and unthinking thoughts? If my assertion that there is no fundamental dualism between mind and body (matter) is correct and if matter is only a manifestation (as it appears to the senses) of the Universal Mind, then the question is how this mind produces (or can have) unthinking thoughts. If the world is a product of a Mind, then its sole nature and purpose must contain the idea of possibility through development and evolution. The material world is only possible through variety in total diversity, universality, and infinity (as a potential). This variety implies order, and this order means hierarchy.

Suppose the Universe is understood, first and above all, like life. This life must have its logic (not necessarily human logic), which feeds this purpose, and that is the preservation of life and its meaning, in its universal connotation, as it is and not as we necessarily see it or would like to see it.

Although everything is not based only on thought but is the thought itself, this still does not mean that everything that came into existence, as we see it, must be aware of its existence or the thought feeding it.

If there is no real matter (other than matter as we perceive it through senses), then even "matter," unaware of itself, is only a manifestation of a mind expressed through something unaware of itself yet serving an essential purpose in the whole structure of the world (as a manifestation).

If the world itself is the Universal Mind's primary purpose, then this world is the only subject, one living organism, to the very Being (Ultimate Mind) that feeds it and sustains it. If it is one organism, it becomes easier to understand why it can be or contain thinking and unthinking "thoughts."

If we analyzed a human organism biologically or in any other way, part by part, without taking it as a whole, we would soon find out that none of these parts, taken separately, would be aware of anything, either of existence or thinking; not even any part of the brain. What is a thinking thing, a thinking thought, a personality, or an I?

We can hardly find anything in the Universe that does not contain information in one way or another, irrespective of its awareness or unawareness of itself, because there is a law to be found everywhere, from atoms to galaxies. These laws are information ("thoughts"). Even if the information is a program, there is still a "thought" powering it.

We would have to separate the thought, as it is usually understood, to understand the thought as an instruction, the way to the way. If we acknowledge thought in this manner as a function of a living mind, then this thought has different levels of manifestation and expression.

If we understand the Universal Mind (Being) this way, we know that the mind (thought) becomes its matter. The mind (thought) is the medium and matter. The mind becomes its material.

Awareness or unawareness hides the purpose of every particular mode (thing) with its specific function and purpose. The purpose of every single mode is not awareness on every level but to serve the higher modes it is a part of. Without these particularities and modes, no awareness is possible in the actual Universe, which means that unaware information is the ultimate source of awareness and that awareness itself is impossible without these (lower) modes (unawareness). However, lower modes are possible without awareness.

Bertrand Russel About *Cogito Ergo Sum* Argument

> "Descartes (1596—1650), the founder of modern philosophy, invented a method which may still be used with profit—the method of systematic doubt. He determined that he would believe nothing which he did not see quite clearly and distinctly as true. Whatever he could bring himself to doubt, he would doubt until he saw reason for not doubting it. By applying this method, he gradually became convinced that the only existence of which he could be quite certain was his own. Descartes imagined a deceitful and powerful demon who presented unreal things to his senses in a perpetual phantasmagoria; it might be improbable that such a demon existed, but still, it was possible, and therefore,

doubt concerning things perceived by the senses was possible.

But doubt concerning his existence was not possible, for no demon could deceive him if he did not exist. If he doubted, he must exist; if he had any experiences, he must exist. Thus, his own existence was an absolute certainty to him. “I think, therefore I am,“ he said (*Cogito, ergo sum*); based on this certainty, he set to work to build up the world of knowledge again that his doubt had left in ruins. By inventing the method of doubt and showing that subjective things are the most certain ones, Descartes performed a great service to philosophy, making him still useful to all students of the subject.

However, some care is needed to use Descartes' argument. 'I think, therefore I am' says rather more than is strictly certain. We are quite sure we are the same person today as we were yesterday, which is undoubtedly true in some sense. But the real Self is as hard to arrive at as the real table and does not seem to have that absolute, convincing certainty that belongs to particular experiences. When I look at my table and see a certain brown color, what is quite certain at once is not 'I am seeing a brown color,' but rather, a brown color is being seen.' This involves something (or somebody) who (or which) sees the brown color, but it does not involve that more or less permanent person we call 'I.' So far as immediate certainty goes, it might be that the something which sees the brown color is quite momentary and not the same as the something which has some different experience the next moment.”[18]

---

[18] Russel, Bertrand. (1912). *The Problems of Philosophy : Chapter II: The Existence of Matter.* pp. 27-9. New York: Henry Holt and Company, London: Williams and Norgate.

Comment:

Regardless of how Descartes formulated his argument, his primary focus was to prove that something is thinking. That something doing the thinking exists irrespective of the idea of yesterday or tomorrow and irrespective of the possible change. At the same point, if something is aware of anything, it means that something exists; otherwise, it would not be possible for it (whatever it may be) to be aware of anything. It does not matter if awareness is right or wrong or what it represents, but awareness is proof of existence.

Still, things that are unaware of themselves may exist. Only the Nothing does not exist. Whatever we can qualify or imagine as something exists. Thinking is proof of itself (the thinking), not the self. The thinking itself is the "self" (thinking "self"), existence irrespective of the uncertainty of personality and the self-awareness or unawareness of certainty or its own identity or delusion about it. The thinking itself *exists*, regardless of the self and who thinks.

The main conclusion is that the self in the *cogito ergo sum* argument is less important than it may look like at first sight. We cannot be sure about the whole concept of reality, not to mention the self. If the entire reality we experience, including energy and mass, is the "program" of the Universal Mind, what can we say about the individual self? The whole purpose of "reality in plurality" is existence, and the "self" (or an idea of self and ego) is the result of existence and not of the self itself. When doubting the self, Descartes' emphasis, although he used the word I, was not literally on the self but on that which thinks, whomever or whatever, at any moment; otherwise, there would be no thinking in the first place. It may only be thinking that thinks. The doubt was not if his self or ego, or his idea about them, was real or imaginary but on thinking as such, irrespective of personality. The individual self can in no way predate existence, regardless of what that existence is. Existence presupposes the self and thinking—be it illusory or not.

The attempts to refute the *cogito ergo sum* argument based on the idea of the self are useless and futile because the primary purpose of the argument was to prove the existence as a phenomenon and not necessarily the particular transitory, or potentially illusory, self. Descartes wanted to show and prove that existence is universally the "self" itself (in a more profound sense) and that it is beyond any doubt, regardless of the sense of self, a particular self, or I. That which provides the basis for the "thinking" or the "self" or the "I" exists, be it reality or illusion, with no difference. There could be no effect caused by something if there were nothing in the first place. That is a *contradictio in adjecto.*

EXISTENCE AND NONEXISTENCE

If we state that there is something, this statement confirms and proves the existence of something (even if it is not a confirmation or proof of any particular something). Otherwise, there would be no way to state anything if there was not something in the first place. Even if there is nothing but an illusion in our head, it is still the existence of an illusion in our head, which is something.

Despite scholasticism or skepticism, skepticism is the confirmation and proof of existence. It follows that if we affirm and confirm the existence, we must, at least, accept the context of existence irrespective of its possible manifestations or variations. We can find this context in the relationship of the Absolute and the Relative and the context of Being and Nonbeing, Something and Nothing.

From the point of existence itself, it is not essential (crucial or philosophically viable) if existence is an illusion or even if the whole existence, in its absolute totality, would be an "illusion," for the existence of existence is self-explanatory and self-confirmatory, even if it were an illusion.

Paradoxically, the bigger problem arises regarding non-existence or nothingness. How do we define Nothingness? What is Nothingness? If there is Nothing, then we affirm or try to confirm Nothingness (Nonbeing), which means that we establish or attempt to confirm the existence of non-existence or nothingness.

If we think in this way, then we indirectly state that non-existence is impossible because the very existence of its nonexistence denies its non-existence. If we continue following this thought, we may conclude that it sounds so paradoxical that our knowledge, comprehension, the world itself, or, perhaps, our language by which we try to explain and understand it is insufficient.

By their nature, abstract theoretical questions lead to tremendous philosophic or scientific discoveries. In that sense, they are essential and should be highly encouraged, regardless of how abstract, impractical, or inconceivable they may seem. On the other hand, if such questions only satisfy the fancies of those pursuing an argument for the sake of an argument, they are mainly worthless.

Philosophy must strive to establish common sense (as Thomas Reid would have liked it) as the highest model for its method. We don't want to say that every easy or comprehensible thought and argument should be the basic criteria. This only means that skepticism has limitations.

If our mind, regardless of being limited, is viewed in the context of reality or existence, then we can assume one existence consisting of all other existences and consider it as one reality of which, most likely, as our experience shows, most parts will not have the complete grasp of its totality. Nevertheless, they will still be able to participate in it, with a larger or smaller amount of knowledge coming from experience (with or without the help of language).

Even if we assume all Reality, all existence, in its totality, as an illusion, we still affirm and confirm the existence of reality—the existence of existence itself. What matters is existence itself and not the attributes we apply to our idea of reality. If there is existence, then there is a reality. Existence itself is a reality. Our characterization of reality as an "illusion" doesn't change the fact that there is reality (even if it is a reality of an illusion). If there is a reality, then there can be no illusion about the existence of existence. The only illusion may come from our understanding (or misunderstanding) of reality and its various manifestations and expressions.

If reality behaves and acts upon us in ways that may seem like an *illusion*, then this, which appears to us as an *illusion*, is not a bit removed from *reality* itself as it is, but only from our understanding of it. Even if we accept illusion as an argument against reality, this is not only a weak argument but also proves what it tries to deny.

The logical question would follow: Is the purpose of *reality* an *illusion*? Or: What would it be if it were not an illusion? Before answering these questions, we would have to define reality and illusion in the best possible way. What is reality? What is illusion? Why can reality not be an illusion to be reality? Would the world be better if everything was literal in an obvious sense? We have to understand that the significant part of reality is its mystery. When reality becomes too real or evident on every level, without mystery, reality loses its purpose and becomes, perhaps, worse than an illusion in our usual sense of the word.

We gradually arrive at the ethical question of better or worse, which must imply which reality is better or worse. Since this automatically becomes an ethical dilemma, it creates an aesthetic dilemma: Which reality is more beautiful? Or, must all reality, irrespective of all its potential and possible variations, unavoidably comply with or conform to some principle of beauty—the Law of Beauty—which must be the basis of harmony if reality inclines toward a comprehensible and intelligible form, order, and purpose.

EXISTENCE AND LIFE

Everything that exists is not aware of its existence. Still, everything that exists is a product of some kind of idea (information, thought). Thinking does not equate to the whole existence but only to the existence of the thinking beings. Nonetheless, non-thinking beings still exist. Lack of thinking does not equate to non-existence. All beings—thinking and non-thinking—exist.

The whole purpose of the Absolute is contained much more in the meaning of existence than in the existence itself, regardless of meaning. The meaning and purpose of existence is the experience of life. Therefore, we cannot apodictically state, as Schopenhauer, that the driving force of the whole Universe is the will (or energy) per any cost but rather the meaning. Not everything is aware of its existence and life, but it serves life as a whole.

Existence, devoid of meaning, is meaningless. Meaning is the ultimate goal of both Absolute and existence. However, meaning, although priceless, is not without a price. If the world had no "price," it would have no meaning. We need some hardship to find meaning in anything. Even when we talk about beauty, we must "earn" it. Only nonexistence is unconditional. Life and existence are conditional and contingent.

If we look at these ideas or phenomena from this perspective, it becomes easier to understand suffering and "evil" in the world. We shall reevaluate every issue and idea. Do we fully understand the roots of evil and good?

Nonexistence equates with death in a way. But, for death, something must be born to be able to die. Non-existence excludes both birth and death. On the other hand, everything that does exist is programmed or destined to motion. Without some movement or growth through space and time, there is no actual existence, but rather the non-existence camouflaged in the robe of "non-existing" existence, which was only a dead existence at the same spot forever, without motion, without time, without birth or real death, which equates non-existence.

Existence is in constant motion and growth (regardless of progress or regress), moving through time and space (including entropy) with all its other attributes. Only something alive can contain and reflect these attributes. Existence, in and of itself, is life itself. The ability and capacity to think do not equate to the wholeness of life. Still, life itself equates to existence as a whole, including the existence of inorganic matter or something we are used to calling "dead matter." Everything that exists, regardless of our conception of it and how it appears to us, is alive. Existence is life.

Beyond existence, there is no death, only non-existence or Non-Being. But without non-existence or Non-Being, the existence of the "material" world would not be possible, and consequently, life would not be possible. Non-existence, or Non-Being, is characterized by an equally important "power" in the existence of existence, or life, itself. (Nonbeing provides or secures the possibility of birth and death or the illusion of transformation.)

Does the word life reflect life itself as a whole or only an idea of what we think life is and not necessarily what it is? Our traditional thinking limited life to organic matter. Our idea about life is different in that we think everything that exists is alive and is life. Once the word life becomes rich enough to demonstrate and represent the wholeness of the Universe and not our original and narrow idea of organic matter, our idea of existence equating to life becomes more apparent. Thanks to this enrichment of the word life, we realize that our usual and traditional usage of the word life, in the strict and narrow sense, must have been partially wrong and limited. But, if the whole reality of existence, including inorganic matter, becomes life when we use the word life not only in the way we used it traditionally, then even something we call "dead" matter becomes alive. However, we did not treat or consider it as alive before.

Even if we proclaim that life equates to existence, the word life, irrespective of its correct and applicable meanings on various levels, is still insufficient to describe or relate to the totality of life. In the terminological sense, we must enrich the word life to mean or include the whole existence, irrespective of our inherited way of thinking. In this way, the new term of the same word, with an enriched meaning, would be rightly established.

The life of non-thinking beings does not equate to the existence of all. The sheer existence of anything is proof of its life; otherwise, it would not exist. Only non-existent is not alive. Life would not be possible if existence did not contain life in its totality. What is manifested as life, appearing to us as real life, is only the evolution or transformation of existence, which is already having a life. Our understanding and description of life do not equate with life itself but only with our definition and understanding.

Existence and Values

Do we try to conclude or come to the truth from the point of the truth itself or from the end of our definition or understanding of any particular phenomenon and the truth itself?

The truth will always be what it is; the facts will always be what they are. Since we cannot change the truth or the facts, we can only change (improve) our minds, descriptions, and understanding. As long as we fight to preserve our truths and facts, provided they are not representations of the "absolute truth" and facts, we fight against the facts and the truth under the disguise of our definitions and proclamations sold as "real truths and facts."

Any attempt to change reality is futile, but attempts to impose our views and definitions onto reality can be profitable. We often try to impose our ideas and descriptions, irrespective of their merits and correspondence to reality and facts. Then, we try to make reality fit our views and definitions instead of the opposite. At this point, truth and facts become irrelevant except our ideas about them (which can easily be a trick), and this is the highest level of dishonesty, intellectual and social. Often, this kind of thinking and this kind of practice lead to dogmatic thought and demagogy, which is a basis for all sorts of oppression and enslavement (of souls), either politically or religiously.

This kind of thinking, in the fields of arts and sciences, leads to different types of manipulation, calculated to lower the standards to promote personal ideas or "qualities" that are not per the highest standards. If that is the case, the only way to establish ("legitimize") qualities or merits of lesser value is by lowering the standards. If everything is possible and has a value, the point of discussion or interest shifts to the question of any particular value of anything, irrespective of its merits. In any field, the supreme qualities and their merits result from the highest standards, not some arbitrary judgment.

(If there is no criterion in terms of the almost absolute hierarchy of values and the nearly absolute value of some facts or truths or near absolute values of some works of art, then it is possible to establish the appearance of similar values or qualities in a work of art, for instance, of lesser value. We can easily find proof of this in some works of modern art.)

The real fight is the fight of facts among the facts, not the fight of ideas among ideas or theories among theories. More precisely, only those ideas and approaches, or only those qualities and values that represent the facts and the truth in the best possible way, have the merits that entitle them, to a larger or lesser extent, to the very same facts or truths, proportionately to their value. Since there is no absolute entitlement to the facts or truths, there can be only the entitlement, to a larger or lesser extent, to the acceptance of the values represented by particular ideas, theories, or works of art, depending on their intrinsic values that depict the facts and truths of existence and life itself in the best possible manner and with the highest level of accuracy or beauty in the arts.

The measure of the quality of ideas or scientific achievements is their accuracy or exactness. The standard of value, or quality, in the field of art is the level of "pure beauty" (including intellectual power, wisdom, and ideas containing justified novelty) represented by the intrinsic value of a particular work of art (in its justifiable novelty). Even beauty is the measure, or one of the possible measures, of the realities and facts of life and existence. Even beauty has its dimensions and exactness. The sheer immeasurability of beauty is the measure of its infinite value. We cannot represent, describe, or define the factual state of beauty in a better way than by beauty itself.

The objective measure of beauty is beauty itself. The real measure of life is life itself. The objective measure of existence is existence itself. We are part of the beauty of the existing world. Through us, the world measures and enjoys itself in its myriad ways. Life and the world impose standards, but we also make and impose our standards. Our standards must be in accordance, to the highest degree, with the factual state of the world to be a real representation of the world at its best. If our standards deviate from this high demand, then what is left is mostly simulacrum, proportionate to the degree of our deviation or departure from the highest possible standards.

The Will to Truth and Goodness

Living by the highest possible standards means opening the doors of the highest truths through the possible means. To live by the sheer desire or "will to power" means to open the labyrinths of personal truths and thoughts that we camouflage as facts and truths themselves. Only the desire and will to goodness can be the model for obtaining the highest truths possible. In that way, the highest obtainable power comes not from the will to power but from the will to truth itself, representing the will to goodness and wisdom obtained in such a manner.

Only the real truth can serve the real purpose. If a desire or will is not represented, first and foremost, by the desire for the truth itself, then it cannot have enough power of merits to deserve real power. If we obtain the actual power, regardless of its merits correlating to the real truth, then this power is not real power but either an abuse of power, an illusion, or a simulacrum.

On the other hand, if the highest possible degree of truth is obtained through the highest degree of genuine desire or will for truth, coupled with personal ability and genius, then the result must lead to some power.

This power is the result of the will to truth, measured only by the truth, irrespective of personal interests, and as such, it equates to goodness. Therefore, goodness is the only source of real power (not actual power sold as power, though it may be abuse, trick, or anything else). The only source of the ability to find the real truths and decipher the actual state of the world is goodness. We can find the essence only in goodness because it is a channel through which the real essences travel (through which wisdom flows).

Only goodness can understand and measure the highest and the deepest truths because it is never satisfied with the sheer power of winning but only with the power of real power, the truth itself, represented in the best possible way. If the measure of power is not accurate, then that is an illusion or abuse of power, which can be measured and felt (unfortunately) by other humans but is not a real measure of real power.

The power of real power would lead to the representation of goodness, obtained by deserved truth (ultimate wisdom). A deserved truth is the source of happiness (through wisdom) and the power without the representation of power in the way we are used to. Real power, as the representation of the truth, is “shy” because its purpose is not to impress, oppress, or enslave but to enliven, enrich, and empower life itself (which is the life of all and not only those who hold power).

Goodness is not a weakness. Weakness sold as goodness is the abuse of goodness and not its representation. Goodness is not without rigidity and is not unconditional in its value; it is only unconditional in its aims, purpose, and service.

The sheer will to power is corrupted not by power itself but by an undeserved power. Undeserved power is only possible through a compromised or corrupt will. Power doesn’t corrupt. Only an undue and excessive power corrupts. If we say that “absolute power corrupts absolutely,” we mean that the insatiable (“absolute”) appetite for power, irrespective of merits, corrupts. But the deserved power or the deserved absolute power, if even possible to imagine, would not be expressed or manifested as such because it would be an actual representation and manifestation of goodness, which is different and opposed to our view of real power.

The rule of real power is almost invisible and incomprehensible. The manifestation of real power is not the power itself or how we perceive and look at power but the consequences of hidden truths and facts and their effects upon us and the world.

EXTREME SKEPTICISM

Extreme skepticism is a negation of itself, for, by its nature, it confirms its existence, which is a confirmation of reality. It also confirms the existence of that which it questions, securing the additional proof of its existence and the existence of its subject (world, reality, illusion, hallucination, or whatever that may be).

The purpose of philosophy is not a sterile questioning of everything and game-playing with everything by those who cannot deal with real subjects but fancy themselves in the pursuit of endless games. The only merit of these games is that nobody can prove some phenomena, at least not yet. Such a state of affairs gives them sufficient reason to additionally question, scrutinize, disbelieve, and suspect because they think there is no proof to contradict them. But they forget that the only merit of such pursuits is a lack of merit. And since there is no merit, the only way to establish merit is by artificially producing questions and supposed philosophical problems, which would, in fact, only serve the proponents of such artificial issues.

Not all questions and problems are real and legitimate philosophical problems, and such sterile pursuits are not genuine and cannot be treated or accepted as genuine philosophical problems except to promote their proponents. We can equate extreme skepticism to parasitism because it questions and undermines that from which it gets its energy and life. Its own life is clear proof of this falsity and paradox.

Every serious endeavor and attempt to understand the world and reality better is philosophy, in a way, and any person pursuing such goals is a philosopher in his or her own right. On the other hand, not all who call themselves philosophers are real philosophers unless they seriously and genuinely resolve outstanding philosophical issues. To doubt for the sake of doubting is not only extreme skepticism but is also sterile.

DETERMINISM AND FREE WILL

According to Daniel Dennett[19] and many other philosophers, free will and determinism are compatible. Despite my agreement with this view, I must admit that my way of coming to the same conclusion may be slightly different, contain different motivations, and come from a different perspective. Therefore, this topic deserves further clarification; although findings may be the same, they may be so for different reasons. Even if they would be the same for the same reasons, they might be based not only on different approaches to the same problem but also on a different perspective in viewing the world—understanding the world, the concept of the world, its potential, and its appearance to us.

Different conclusions to which Pierre Simon Laplace (*Philosophical Essays on Probabilities*[20] [1814[21]]) arrived stem from almost the same subject (the world) analyzed by Dennett. We must credit Laplace (which Dennett did) for thinking about the same problem two centuries ago without possibly being affected by the discoveries to which Dennett and other philosophers and scientists were exposed. However, we must emphasize that some other philosophers and scientists before Laplace treated the same

[19] *Lecture on Free Will*, Edinburgh University

[20] Laplace, Pierre Simon, *A Philosophical Essay on Probabilities*, translated into English from the French 6th ed. by Truscott, F.W. and Emory, F.L., Dover Publications (New York, 1951) p.4.

[21] Laplace mentioned this idea for the first time in 1773.

subject, including Baron d'Holbach and Roger Boscovich (*Ruđer Josip Bošković*) in his *Theory of Natural Philosophy*[22].

> "Laplace's Damon" (argument):
> "An intellect that at any given moment knew all the forces that animate Nature and the mutual positions of the beings that comprise it, if this intellect were vast enough to submit its data to analysis, would condense into a single formula the movement of the greatest bodies of the universe and that of the lightest atom; for such an intellect nothing could be uncertain; and the future, just like the past, would be present before its eyes."

— Pierre Simon Laplace, *A Philosophical Essay on Probabilities*

There is nothing wrong with this argument since it is only hypothetical in terms of "An intellect which at any given moment knew all the forces that animate Nature ..." This is not a positive or negative statement about determinism but only an intellectual proposition or question of what the case would be if there were such a "vast enough" intellect. Another question is if Laplace's own belief or faith would lead him or not to such a conclusion. He only states that "an intellect which at any given moment knew all the forces that animate nature ..." which is not proof that such an intellect exists or that he unconditionally believes in such an intellect. The mere intellectual proposition about an imagined intellect (not necessarily a real one) under the proposed conditions (not necessarily the real ones), we shall treat only as a hypothetical question or proposition or statement and not as an apparent belief (though there may be a clear belief behind it). Furthermore, this proposition doesn't prove how it would undermine the

[22] *Theoria philosophiae naturalis*, 1758

compatibility between determinism and free will even if such an intellect existed.

Laplace's conclusion under the proposed conditions is proper and must be true. But the question is not whether the conclusion itself is true if the argument, Laplace's Damon (actually intelligence), does not represent (demonstrate) or prove the fact (truth) but only a possibility that this may be a fact (if such an intellect existed). We cannot say that this is a definition of determinism by Laplace but a possible vision (of a definition) of a universe under the proposed conditions.

To prove this to be a definition, not only of determinism but also of the actual world (Universe), to which goal Laplace aims, Laplace would have (or should have) offered "proof" of the actual state of the Universe. Since this was impossible, he defined a world under such "ideal" conditions through an example as a proposition. The question of whether the world, under such "ideal" conditions, would be ideal is highly debatable (we would dare to think that such a world would be not only far from perfect but on the verge of horror). From this argument, it follows that if such a world were not, in fact, ideal, then imagined or proposed "ideal conditions" would not be ideal either.

Determinism implicitly seeks perfection, but the idea of determinism pursuing perfection is the pursuit of its own perfection (rather than perfection outside itself). But we would first need to know what perfection and an ideal are. It would not matter much if, by analyzing this, we conclude that the ideal and the perfect are the same, in which case we would use these words and terms interchangeably. But the problem is how to define the basis (on which determinism is, and must be, largely based) of the whole universe if we do not understand exactly the meaning, or all possible meanings, of the terms perfect and ideal, not to mention our understanding or misunderstanding of the actual state of the Universe.

Therefore, we must conclude that any proposition (definition) of abstract concepts (linguistically or philosophically) is not necessarily a proposition (definition) of that to which these concepts or propositions (definitions) ought to be applied. This kind of thinking, or definitions coming from such thinking, is inherently false.

On the other hand, if the supposed or proposed argument reflected reality, it would be a definition of a fact and not a proposition ("definition") of an imagined concept. Laplace didn't show interest in proving that this is the actual reality or even stating it, so it must be that he either imagined his argument as a hypothesis or he believed that this example argument contains enough merits of itself to justify not only the argument but also the fact or truth itself.

From our point of view and the idea of the Universal Mind as having a potential for infinite variety, we can conclude that even an "intellect" of a Universal Mind cannot predict and know all the varieties of its potential and the potential of all possible universes in infinity, which saves it from its own perfection while opening the door for infinite potential of the "free will."

DETERMINISM AND PROBABILITY

We may understand determinism as “the thesis that there is at any instant exactly one physically possible future” (Van Inwagen, 1983, used by Dennet in his Edinburgh University lecture). But, a theory, or a thesis, must be based on our concepts, knowledge, and understanding of reality. Our very concepts, understanding, and conclusions about reality may be only partially correct or accidental since there is not enough knowledge (not to mention evidence) to support the particular theory or thesis.

In light of these ideas, it would not only be more economical but also prudent and wise to consider the reality out and beyond the customary spectrum and, by doing this, to come to conclusions or theories that would more significantly, possibly more accurately, reflect not only our ideas but the reality itself as it could be, beyond established and accepted paradigms, and, ideally, reflect the reality as it is. In that way, we would provide the ground (or possible grounds, or open the possibility) for defining the actual reality in a more tangible or, ideally, more accurate and useful manner than defining ideas and concepts based on theories, which may very well be, more or less, futile attempts to define the indefinable, depending on the degree of accuracy of any single approach.

Dennett uses the example of quantum physics to prove that “at any instant, there are many possible futures, and they are completely undetermined.” Once the organism of the Universe is in motion, its destiny is determined to a large extent. This destiny will be different if the same Universe starts motion (walk, experience, life) at any other “moment.” Every destiny of every new Universe or revival is different. That is the potential of the Absolute. This potential or its variations are not known even by the omnipotent intellect or God. Even an “intellect, which knew all the forces that animate Nature at any given moment,” could not know precisely how its potential “materialized” in every new birth, revival, and cycle.

INFINITY IS IN THE INFINITE REBIRTHS OF THE UNIVERSE

Every possible universe is different from all the rest. No new or reborn universe will be the same as the one before. The accidental nature of these rebirths shows how free will functions from the beginning of anything. It also shows that the world, which doesn't only copy itself, is more interesting. The immaterial Primordial Being transforms itself in countless, infinite ways, allowing even the functioning and possible satisfaction of ethical laws on the absolute level. On the level of its potential, we may say that the world is "perfect," yet we cannot foresee this perfection. Potentially, everything is part of everything else, and we can strive for the highest possible perfection in due "time." The pain of one is the pain of all; the guilt of one is the guilt of all; the king is a bum at the same time. Every element makes the World and is the World. Everything is God, and the World is God. In this sense, everybody and everything is God. No bill will be unpaid. Justice and ethics of the Universe go beyond the human, Earthly concept of justice and ethics. Even Nietzsche's idea of eternal recurrence, if it existed in reality, would not negatively affect the potential of the Absolute because of its absolute capacity to absorb, transform, and integrate all forms of life into its new manifestations and cycles.

CHAPTER 7

WHAT IS GOD?
CONCEPT OF GOD-WORLD

God and the idea of God have obsessed philosophers since the beginning of philosophy. For any serious philosophical inquiry, it is essential to establish precise prerequisites and rules if we want to get the desired results or get close to them. If the subjects or the terms we use are not precisely defined, we may shift in different directions, not knowing that the result we get is different from what we wanted, or we may move, more or less, within the realm of logical fallacies.

One of these ideas and terms is God. But before we start pondering about God, we have to set up the philosophical, logical, and linguistic frame within which we investigate the idea of God. If we do not define our ideas about God, we can hardly accomplish what we want, regardless of our desires. Our ideas about God are not God. Our ideas about God cannot create God. Our ideas about God are just that—ideas.

In religions, the idea often takes the subject's place. The idea becomes the subject itself. Since this is impossible, human words are used as the words of "God," which serves as "proof" that God said them. In this way, human words, sold as God's words, cannot be questioned and are "proof" of God's existence. That's the idea that takes the place of the subject.

Everything would be easier if this were the only problem relating to God and philosophical inquiries about God. The problem is much more complicated because wherever we turn, we either find logical fallacies, misusage of language, or inadequate comparisons to try to make a statement, to try to oppose and deny the other side's argument and prove the "truth," which usually turns out only to be our "truth," our view and not the objective truth. We can find this among believers, atheists, and agnostics when the idea avoids or bypasses the subject in a real sense.

Since God is not a subject of which we can have tangible evidence through direct experience, or at least, we think there is no clear and direct experience, and we are mainly dealing with our ideas about God and not God himself.[23] Based on our thoughts, God can be many different things to different people. In Judeo-Christian, theistic tradition, God is the creator of the world from nothing—*creatio ex nihilo*. In this view, God is not its creation.

When we ask somebody if they believe in God, we expect them to believe or not in what we think our idea of God is. But what if God is not what we think it is? Many believe that some undetectable higher Force they do not understand governs the Universe. This kind of belief is more rational because there is something they believe in, but religious books do not describe it.

God is not what our idea of "It" is. There is no religious book that offers a satisfying picture of God. Nevertheless, if we think of God as a force, even atheists may agree (if not now, then perhaps in the future) that there is a unifying force in everything. That force is the Absolute, whereas the Being responsible for creating motion, energy, and "material world" is the force we may call God, the Ultimate Being, Ultimate Force, or Universal Mind (Spirit). Still,

---

[23] No pronoun can rightly fit or express the idea of God. We use pronoun *himself* only as a conditional "personification" although it is impossible to personify God in a true sense since God is Everything except Nothingness. Nevertheless, at its highest point, it becomes Nothingness.

the world is not only the product of its creator but is the creator itself in a different mode of existence.

For Milesian philosophers Thales (c. 626/623—c. 548/545 BC), Anaximander (c. 610—c. 546 BC), and Anaximenes (c. 586/585—c. 526/525 BC) there was an ultimate principle they called *arche*. For Thales, this ultimate principle from which everything originated was *water*; for Anaximenes, it was *air*; and for Anaximander, it was *Apeiron* (limitless), whereas, for the Pythagoreans, the number was the ultimate principle.

For Heraclitus (c. 540—c. 480), *arche* was *fire* from which everything originated, but *Logos* was the ultimate principle uniting everything and connecting opposites.

For Anaxagora (c. 500—c. 428 BC), a hundred years after the Milesians, the ultimate principle was the *mind* (*nous*), which is limitless because it is not material.

According to Plato (428/427 or 424/423—348 BC), God is a perfect Being, eternal itself, shaping the universe through eternal forms (*eide* [form: *eidos*]). The imperfections in the world result from the material nature of the universe.

According to Aristotle (384—322 BC), God is an *unmoved mover*, a primary cause, and a possessor of a divine mind (*nous*).

According to Plotinus (c. 204/5—270 CE), God is Intelligence or Mind (*Nous*), and the world is created out of God (*ex deo*) and not from nothing (*ex nihilo*). "The power of the One is to provide a foundation (*arkhe*) and location (*topos*) for all existents. The foundation provided by the One is *intelligence*. The location in which the cosmos takes objective shape and determinate, physical, form is the *soul*." … "The being of intelligence is thought, and the thought of intelligence is Being." … "No idea is different from intelligence but is itself intelligence." Plotinus accepted the Stoic's idea of *logoi spermatikoi;* for him, *logoi spermatikoi* is a bridge between the soul and the material.

According to St. Augustine of Hippo (354—430), "The highest good, than which there is no higher, is God … And consequently, if He alone is unchangeable, all things that He has made, because He has made them out of nothing, are changeable."[24] Augustine also used the idea of *logoi spermatikoi* in the context of seminal reasons (*rationes seminales*, Latin from the Greek *λόγοι σπερματικοὶ* or *logoi spermatikoi*), or "seedlike principles," "causal "principles." Based on this theory, God created the world by inseminating the void with seed. Other Christian thinkers accepted the idea, including Justin Martyr (100—165), Athenagoras of Athens (133—190), Tertullian (155—220), Gregory of Nyssa (335—395), Bonaventure (1221—1274), Albertus Magnus (1200—1280), and Roger Bacon (1219/20—1292).

According to St. Anselm (1033/4—1109), God is a "being than which no greater can be conceived." As we can see here, this is a reformulated idea of St. Augustine, who thought similarly centuries earlier, not to mention ancient Greek philosophers.

Thomas Aquinas (1225—1274) states that God is a simple being. Although God is eternal, a material world, Universe, is not eternal. For Aquinas, God's existence is his essence, the basis of Divine simplicity. For anything else, there is a distinction between existence and essence. Aquinas defined his five arguments for the existence of God in his book *Summa Theologica:*

1. The First Way: Motion. (The argument from "first mover.")
2. The Second Way: Efficient Cause. (The argument from universal causation.)
3. The Third Way: Possibility and Necessity. (The argument from contingency).
4. The Fourth Way: Gradation. (The argument from degree.)
5. The Fifth Way: Design. (The argument from final cause or ends [Teleological argument].)

Spinoza's (1632—1677) *Ethics* starts with a clear framework, explanation, and definition of his terms. In that way, the

[24] *On the Nature of Good.*

philosophical inquiry becomes more accessible and precise for a reader or interpreter to understand and grasp. When Spinoza, in his definitions, uses the term substance, we understand that it is God. But when the term substance reappears under point III and then again under VI, which treats God, we must question why. For Spinoza, there is substance and substance. What is the difference between the substance under III and VI? We would say that, according to Spinoza, the ultimate, infinite substance is God, and everything formed is of the same substance. If that is the case, all substance is God or Nature. If all substance is God, then the question is, why separate substance from substance?

Spinoza wanted to highlight the difference between the infinite substance of the ultimate Being, God, and the substance that makes Nature in all its forms. But nature, or anything in nature, is substance "which is in itself and is conceived through itself and does not need another "thing" to form it." Nature is just a manifestation or mode of God or Substance.

Substance (*substantia*) is not a new term and has been used since Aristotle, if not earlier. Perhaps the substance is interchangeable with terms like *arche,* aether .... fifth element, *proton archon* (first principle), Plotinus' Divine mind (*nous*), or intelligence. Here are Spinoza's definitions:[25]

Of God

DEFINITIONS

I. *By cause of itself I understand that whose essence involves existence, or that whose nature cannot be conceived unless existing.*

II. *That thing is called finite in its own kind (in suo genere) which can be limited by another thing of the same nature. For example, a body is called finite because we always conceive another which is greater. So a thought*

[25] *Ethics.*

is limited by another thought; but a body is not limited by a thought, nor a thought by a body.

III. By substance I understand that which is in itself and is conceived through itself; in other words, that the conception of which does not need, the conception of another thing from which it must be formed.

IV. By attribute I understand that which the intellect perceives of substance as constituting its essence.

V. By mode I understand the modifications of substance, or that which is in another thing through which also it is conceived.

VI. By God I understand Being absolutely infinite, that is to say, substance consisting of infinite attributes, each one of which expresses eternal and infinite essence.

STOLEN GOD AND THE RIGHT TO TRUTH

God in the major monotheistic religious books is a "stolen" God. This way, the church fathers tried to establish the right to God and the truth. They told us we have to abide by the words of God they received directly from Him. We never obtained proof, nor can we question the validity of this "truth;" we just have to accept it at face value and take their word for granted. However, God is Truth, and the truth belongs to everybody equally to the degree of availability.

Nevertheless, there can be no right to truth except recognition and knowledge of truth. Once we try to establish the right to truth and God, we leave the territory of truth and enter the domain of demagogy and deceit. The truth is an almost absolute category, and the same applies to God. What is indisputable or nearly absolute can only be partially known. Still, major monotheistic religions not only steal God, but they steal him (It) absolutely, not somewhat.

God is, first and foremost, an idea. Nobody has the right to God, to claim God, even though people throughout ages claimed it, and probably always will through religion. But this God is only a stolen God, made in the human's image (the image of the human race), wearing the colors of his creators and not his own. This God wears not the colors of truth but of deceit. Deceitful simplicity is neither simplicity, truth, nor God. A stolen God is not God at all because it is, above all, an instrument of a human being in the service of a human being and not of God.

We must reevaluate and enrich the idea of God to encompass all inquiries, including scientific and philosophical, to fit reality and the truth instead of serving our more or less sophisticated but still poor or poorly presented and imposed ideas. This enriched idea of God is not only the ultimate goal but an ultimate reality we shall strive to understand fully or get closer to.

God belongs equally to believers and atheists, but believers "authorized" the right to the word God and its meaning. Agnostics and atheists got caught on it, although they would like to know the truth, too. Since God is the truth, atheists did not realize this yet. Once they realize it, they will become believers because they believe in truth and God is Truth. However, this does not mean that the Christian God or the God of Islam is the truth, but that the truth itself is God. The majority of believers, atheists, and agnostics would instead like to know or accept the truth rather than a stolen idea of God and lies (although many of them conceived with good intentions) presented as truth.

Atheists got so trapped that they forgot that God is not necessarily what religious books say It is. We have to redefine the word God to encompass everything. If God is the Creator of everything, then God is Everything, and if God is Everything, then God is the truth, and why not believe in truth or refuse the truth in the name of whatever partial truth? The truth does not care about human-biased decisions and their fights or who first claimed the truth.

GOD AND DEVIL

Regardless of whether the world is predetermined, there can be no Devil. The Devil is not only the antithesis of God but is also impossible. The idea of the Devil as the result of chance, free will, rebellion, or sin is also impossible. The only possibility, philosophically, is that God is the Devil. The Devil, as understood in religions, was invented to cover up all unjustifiable evils, all misunderstandings, all sufferings, and all "sins" of human beings. The Devil is a scapegoat of religions. All that is bad we can attribute to the Devil, and all that is good to God. The invention of the Devil is the biggest fraud of all.

Why would God create the Devil in the first place? If free will is the excuse, why would the omnipotent God allow the Devil to rebel, and why "he" would not predict this rebellion? Why would God allow Eve to sin? Are animals capable of sinning? What is sinning? Is suffering possible only within the realm of human beings? Although animals do not possess or understand the morality of humans, they can suffer. Even if they cannot suffer all human suffering, they can still suffer pain. Does God predict the pain and suffering throughout the animal kingdom? If not, why not? Why would this suffering or distress of animals be less sympathetic to God than human suffering? Who determined that humans are not animals, and is this definition, or generally accepted view, only arbitrary?

No main monotheistic religion offers satisfying answers to any of these questions. No religion is concerned with the well-being of the animal kingdom,[26] and no religion proposes or establishes sin in the animal kingdom. Religious books strictly reserve sin for human beings. No religion offers or predicts either hell or paradise for the animal kingdom, only for human beings. No religion even tries to explain, in a reasonable manner, the absurdity of eternal punishment and its "ethics." What horror can equal the monstrosity of eternal punishment, eternal hell? There is no bigger pain, no bigger monstrosity, no bigger immorality, and no bigger cruelty than the eternal punishment, eternal hell. And only an imperfect, mortal human deserves this eternal measure, eternal punishment. How monstrous is this?

Where is the logic behind the idea that something mortal, fickle, imperfect, weak, could deserve such eternal measure, eternal punishment in its absolute force, which would be too harsh even for the Devil if it existed? Only absolute crime deserves absolute punishment, and Hell, being eternal, is the absolute punishment for its inhabitants. But what possible crime deserves absolute punishment except if a man was capable of killing God, which would be the absolute crime, deserving absolute punishment and Hell? In that case, there would be no God to execute the punishment. Therefore, absolute crime is impossible. Even if God survived, this omnipotent and benevolent God would still have mercy over such a weak and wicked creature as a human being, to exercise his grace and absolute power instead of revenge and save the soul of a lost human being from eternal pain.

Where can we find anything so cruel invented by the Devil in any religious book that would equal the evil majesty of the last judgment, eternal hell, and eternal punishment? It is hard to imagine or invent anything more powerful, more painful, and cruel than hell. It is also hard to imagine that the evil Devil would be

[26] We are simplifying the matter regardless of cow reverence in Hinduism (or similar phenomena anywhere), to mention one example.

able and capable of creating anything as monstrous as the Hell "invented by God," not the Devil, according to the scriptures. Where is the difference, then, between God and the Devil? And if there is any difference, by what criteria is the evil Devil more "malevolent" than the "benevolent" God, based on what we read? If we correctly *read* and make common sense conclusions, there is no more immense evil than the Hell invented by the benevolent God.

But is this possible? Is it possible that a benevolent God would punish eternally weak and wicked creatures as human beings are, and probably must be, based on their limited powers? We answer that it is impossible. A benevolent God would not do all this in this world or the afterlife. It is inconceivable that the omnipotent being, characterized by mercy and grace, among other things, would be so cruel. The all-powerful, benevolent, and all-knowing God would, at least to a certain extent, save human beings from their sins and his "future son" from suffering if all this was a matter of free will and the original sin. All this sinning and suffering could have been easily prevented by the omnipotent and benevolent God without the infringement on free will, without the possibility for sin and eternal hell for the millions of people in the "afterlife life."

The main problem is not God or the Devil but a human being incapable of resolving affairs without "sinning." The biggest source of continuous sinning and misunderstanding is not God's word but a word uttered by a human being who sold it as God's word. These and similar deceits throughout history are the highest forms of the enslavement of human beings, the highest forms of insincerity, and the highest forms of disrespect to a benevolent human being. Skillful and often evil human beings act (enjoy acting), using a benevolent God as the means for malevolent human causes. Even if the reasons were benevolent and often based on ethics, the ends did not justify the means. To trade the souls of millions of people in this life, not in an afterlife, to accomplish either political goals or to provide grounds for easier rule, "ethics" and "legislature" are the best examples of evil will that can be best described as inspired by the "Devil" itself.

The omnipotent God, regardless of free will, would predict the Devil himself and save not the human race but himself from this possibility, which undermines his omnipotence. No Devil can exist as a creation of God unless intended by God. If God did not intend the Devil, then it either does not exist or is God himself. Nothing in the universe, even evil, can be ascribed to anything else except God. No Devil can accept the world's sins, which are God's sins. No Devil can sacrifice himself for the sins of the world or human beings, which are ultimately God's sins. If the Devil would be offered, or serve as an excuse and scapegoat, for the world's sins or human sins, which are, ultimately, God's, then the Devil's sacrifice would exceed in nobleness God's benevolence and again prove that the Devil is a higher Being who must pay a higher price than God for the existence of all.

This dichotomy is not only artificial but is also insincere. God is even what we hate in him. There is more goodness in the Devil than there could be evil in God. God and the Devil are the same. Both are words that describe the same thing. The act of creating is more devilish than anything else. Maybe the creation itself is the primordial sin for which we all pay the price. Maybe Eve did not commit the original sin but the Devil-God himself. Maybe God-Devil himself wanted the fruit of knowledge, the fruit of creation. Perhaps he challenged the absolute peace of the Absolute, the absolute peace of nothingness. Maybe the original sin stems from this provocation of the Absolute Being. Perhaps the absolute emptiness, the ultimate *Garden of Eden*, was disturbed by this interference, and God threw a challenge into the face of nothingness and the face of the Absolute. And maybe for this sin (God's rebellion), the Absolute Father, the Absolute Being, had to expel God from the absolute garden of peace and Paradise into the Hell of Existence, the Hell of Life.

Who is the first rebel, the first warrior, the first sinner? Who disturbed the first and universal peace of the Absolute? God himself. Who pays the price of the original sin of God himself, not Adam's sin? The world itself. The Devil is all we don't recognize in ourselves. God is all we don't understand. But what we don't understand, we ascribe to the Devil. Evil is what we don't recognize, and what we don't recognize, we don't understand. If God is the Creator, his only legitimate son is the World (Christ can only be a grandson, among myriad others in the Universe). But God and the World are the same. Since God is the World, through the World, God creates and recreates himself—the God-World.

Existence is the noblest task for the noblest being. There is no real existence without real life. Real life is impossible in absolute comfort and perfection. Absolute perfection and comfort in real life, or of real life, would undermine its value. The value of life is manifested, among other things, through the efforts of every single being in a universal scheme of things. Without effort and action, there would be no real beauty, and even if there were real beauty, this beauty would be less beautiful because we would take it for granted; it would be too easy and, therefore, less enjoyable.

The highest beauty is life itself. The measure of life is not suffering but existing. The purpose of existence is life. The measure of existing is not absolute comfort but absolute strife for absolute beauty, absolute perfection. The hardship of life, even innumerable suffering, is proof of life as it is and not as it should be, according to us. We can only dream of producing a more perfect existence, a more perfect life, a more perfect eye or jaw, but we are incapable of making life. In this sense, we are incapable of judging, in absolute terms, what good or bad design is, and we are incapable of escaping a given reality, which we may improve and approve of but in no way disprove.

Even if we create a better human eye (biologically), a better leg, a better heart, and a better jaw, this will prove nothing except our drive to continue and improve ourselves and live within the given frame. An idea of creation in terms of fundamental creation of what we see as the Universe (or anything in it) will always be beyond human reach. Whatever we create can only be the modification or mutation of whatever exists. We can explore our potential for novelty within scientific discoveries and arts and not in the realm of primordial original creation beyond human comprehension and reach.

Heaven and Hell

We can turn the picture of the world to mean that hell is not in the other world but here. We can further turn it around to suggest that if there is any punishment in hell, then this world is the place of that punishment where we pay the price, not for previous sins but the price of life, the experience of life. Our very existence is the "primordial sin," and we are sinless sinners.

WHY THE WORLD IS NOT PERFECT?

The argument from poor design in its strict and simplest sense means that the World (Universe) and everything in it is not perfect. To better understand and answer this argument, it is essential to know that the world and everything in it were neither made nor came into being to be perfect as we understand perfection. With the more "perfect" design (to use this term for the argument), conditionally speaking, evolution would either not be possible or would be limited in proportion to its potential.

The real question should be—what is perfection? Suppose there is only the Universe and the idea of the Universe-World as atheists would see it. In that case, the answer must explain, possibly prove, what the absolute perfection of the Universe is, or should be, concerning its potential.

The perfect world is *contradictio in adjecto.* A perfect world is not a world anymore. A perfect world is a dead world. The perfect world contradicts existence. A perfect world, as a multitude, is impossible. A plurality implies meaning and cannot and should not be perfect. Perfect can only be One, but the perfect One is equal to its opposite, which is nothingness. The world, in its plurality, equals life. "Imperfection" secures meaning and purpose in life. In a perfect world, there would be no purpose and meaning. Perfection is only the goal and not the way or mode of the Universe. By becoming perfect, the world becomes nothing since only nothing is perfect. If the Universe were perfect, it would not be the world as we see it but "God" itself. Evolution (as atheists see it) would not be possible in the perfect world because such a world would already be perfect, which would imply that the highest point of "evolution" had already been achieved. Absolute perfection negates any evolution, except if we imagine *retrolution* (to coin a word), moving backward.

Creation of the world does not necessarily imply the all too simplistic idea of design (poor or otherwise) because evolution is not entirely compatible with determinism, which means that only in a predetermined world perfect intelligent design (as defined in religions) would be possible (which is impossible in reality).

The idea of evolution, proving itself before our eyes at every moment, is even less compatible with the simplistic notion of design. Anything designed is static in a way. On the other hand, if design predicts evolution, then evolution must be a crucial part of the design, but what kind of evolution? If design predetermines evolution itself, then evolution must develop as planned and only in one way; otherwise, there would be no possibility for chance or free will.

Most likely, the world was not predetermined to evolve deterministically but following its unlimited, infinite potential. The world's potential contains infinity. It is the potential and capacity of the world for variety through chance. Without chance, conditionally speaking, there would be no free will, among other things.

To celebrate the world would be the most honorable task. If the World is the work of God or emanation of God, this celebration would be the celebration of both the World and God at the same time. This God is not the God from religious books. No book can claim God, and no people hiding behind holy books can be God's representatives on Earth and be its sole interpreters and definers. The church needs thorough transformation. The church's task is not to claim or steal God but to promote and celebrate life in its purest form.

COMPATIBILITY OF FAITH AND ATHEISM

Einstein wrote that “science without religion is lame, religion without science is blind,”[27] but we have to remember that Einstein talks about the “cosmic religious feeling,”[28] which is very different from the religious feeling exercised within the official religions. He explains, “[h]ow can a cosmic religious feeling be communicated from one person to another if it can give rise to no definite notion of a God and no theology? In my view, art and science's most important function is to awaken this feeling and keep it alive in those who are receptive to it.”[29] … “I maintain that the cosmic religious feeling is the strongest and noblest motive for scientific research.”[30] … “A contemporary has said, not unjustly, that in this materialistic age of ours, the serious scientific workers are the only profoundly religious people.”[31] … “You will hardly

---

[27] Einstein, Albert. “About Religion’ (“Science and Religion,” Part II). *Ideas and Opinions*. Ed. Carl Seelig. New York: Modern Library, 1994. p. 46. (Part II from *Science, Philosophy and Religion*, A Symposium, published by the Conference on Science, Philosophy and Religion in Their Relation to the Democratic Way of Lite, Inc., New York, 1941.)

[28] Einstein, Albert. “About Religion.” (“Religion and Science”). *Ideas and Opinions*. Ed. Carl Seelig. New York: Modern Library, 1994. p. 38. (Written originally for the New York Times Magazine. Appeared there November 9,1930 [Pp.l—4]. The German text was published in the Berliner Tageblatt, November 11, 1930.)

[29] Ibid., p. 38.

[30] Ibid., p. 39.

[31] Ibid., p. 40.

find one among the profounder sort of scientific mind without a religious feeling of his own. But it is different from the religiosity of the naïve man."[32]

If we accept that intelligent design is not necessarily what we think it is, or if we enrich the word (term) design to contain additional meaning in extraordinary contexts or ideas about God and creation, we may understand that design may be the design without designing, that the creation may be creation without creating in the way we see it and understand it. We admit that there can be no design without designing and no creation without creating, but what is to be created or designed in the absolute? If the Absolute is the "highest" form of "existence," then it must be, at the same time, absolute perfection. If there is absolute perfection, what creation can match the existing perfection? We must agree that no possible outcome of the Absolute can be more perfect than the Absolute itself. Absolute itself is perfection; otherwise, it would not be absolute.

Why creation, then? Why creating? Is it not better to be perfect and absolute at the pinnacle of itself? We think it is not, as we tried to prove in the preceding pages, and it is not precisely because of perfection and absoluteness. The absolute perfection of the absolute is its death penalty, as we have already stated. Absolute of itself and for itself (without plurality) is nothing.

If the creation of the world needed to be perfect (absolutely perfect), then why create (design) to arrive at the same point of the Absolute, which would be pure idleness, an idle walk from one perfection to another or from nothing to another nothing? If the purpose of creating is not only in creating but more in existing, and not only in existing *per se* but in meaningful and purposeful existing, then this creating or creation must first and foremost accommodate its purpose rather than perfection itself. Its

[32] Ibid., p. 40. "The Religious Spirit of Science." (Mein Weltbild, *Amsterdam: Querido Verlag*, 1934.)

perfection must be valued and measured by and through its purpose and meaning rather than by pure perfection. Measure through purpose makes us see the deeper importance and meaning in the lack of ordinary perfection. Perfection and absolute goodness (benevolence) are not necessarily as we see them and would like to see them. When we measure ideals through purpose, the idea, function, and manifestation of perfection and purpose become more pronounced, although subtly.

To be perfect in the absolute mode, in the same manner, as in the modes manifested through plurality, is impossible. We are not talking about degrees of perfection but about potential, possibilities, and, ultimately, the *sine qua non* of existence in the light of only one possible imperative—the imperative of existence and life.

If absolute perfection is an idle walk from one point to another, which is no walk at all, then there is no existence and no life. If we come from this postulate and premise, the existence and life, as we see it, must be, conditionally speaking, "less perfect." This less-perfect world (existence) is possible only through creation or recreation ("design"). But this creation is not possible without recreation. There is nothing to start from except the Absolute itself—Being and Nonbeing (Something and Nothing). There is nothing to hope for outside this realm. Nothing can be created from nothing if the Being does not create it. Regardless of how omnipotent it is, the Being cannot create anything except out of itself. Even creation out of itself would not be the real creation but recreation because the created being would still be the same (although modified).

Now we come to the idea of *ex nihilo*. We do not know anything about the world before the Big Bang, which could mean that the world came from nothing. But if we ask religious people, nothing can be created out of nothing. But if nothing comes out of nothing, the creation must be the "Child" of existing something, which must be God. If even God cannot create something out of nothing, it must create the world (universe) out of itself. If this creation is the creation out of itself, then it cannot inherently be different from the so-called "creator." If it cannot be inherently different from the creator, then this creation cannot be precisely called creation but recreation. But even this recreation is impossible without the Nothing. In this sense, there is a creation (partially) out of nothing (*ex nihilo*) because the real creating force is the Absolute and not only a God or a Universal Mind. The Universal Mind creates and procreates with the help of Nothingness and not solely from nothingness because *ex nihilo nihil fit*—from nothing, nothing comes to be.

If God is perfect, from a religious point of view, why would creation be needed? Why create a world which is inherently less perfect than the creator itself? Why not create Paradise immediately and avoid original sin and eternal torment in Hell? Paradise is a more peaceful and ideal place. Why challenge and tempt a human being so much? Why provoke a potential rebellion by the "Devil?" Is it the boredom of God? Is it creating solely for creating? Or is it the creation in the pursuit of purpose and meaning? What higher purpose and meaning is there if not perfection itself? Following our already elaborated ideas, we should not view this purpose from the point of perfection. Still, the point of existence and life, for even a perfect God, is dead and nonexistent at absolute perfection. Absolute perfection obliterates existence.

"Perfection" of existence is in its plurality and its "imperfection." Existence contains the whole purpose in a constant, eternal pursuit of perfection. In this pursuit, the mode, motives, and, above all, meaning are found. Even if there were a possibility for perfect particularities within the realm of existence (world, universe), this perfection would still undermine the higher perfection achieved through evolution. The perfection of particularities would undermine the absolute potential of the Absolute for infinity. The world must be "perfect" (or "imperfect") not only concerning the Absolute and not only concerning the particular realities but also proportionately perfect (or imperfect) concerning its potential and the potential of the Absolute concerning eternity and infinity.

The Absolute can exercise its potential only through non-accidental accidents or chances. Chance, although almost accidental, is not accidental. In a predetermined world, chances would not be possible. Even if chances existed, they would only be appearances of chances and not real chances. In a world (universe) established based on its full potential, chance is compatible with determinism. This almost absurd statement becomes logical if observed from the point of view of the potential of the Absolute, which is predetermined. Absolute is not only absolute but also capable of infinity by exercising its infinite potential.

Although the absolute potential of the Absolute and its existence are predetermined, any particular existence is developed, among other things, through chance. The voyage toward infinity allows free will and secures its compatibility with determinism based on the absolute potential of the Absolute itself. The absolute potential would not be possible without the magic force of chance. If a world were predetermined, that would be like playing out the story with the known outcome. In such a world, even if existence, or multiple creations, would go *ad infinitum*, the outcome of any possible existence (universe) would be predetermined and therefore known at the moment of creation. If this were possible, even theoretically, what would be the purpose of such existence (existences) if the outcome is already known? If the sole purpose were to exist, this possibility would not satisfy the ultimate purpose, which is the meaning itself.

The only real purpose and the only real meaning lies in freedom, and there is no real freedom without free will. Free will can only exist if there is an element of chance. But this chance is not a chance as we ordinarily view it, but a chance for possibility and existence, a chance of existence itself. In this manner, the fundamental forces of the world are manifested in the world, regardless of the level of awareness of this knowledge. Knowledge of the world is intrinsic to the world, irrespective of its level or degree of awareness. On the other hand, Chance is the real potential of the world.

Only emptiness and nothingness can provide space to the world; chance is the uniting force of the Being and the Nonbeing. If we view evolution in this context, evolution, as selection, is no longer a random selection or Herbert Spencer's "survival of the fittest" but the survival of existence itself. Whatever survives is thanks not only through combinations and recombination of some otherwise self-organized dead matter, self-powered peculiarly through an infinite series of accidents, but rather through an infinite series of predetermined chances. Determinism is based more on chance than on determination. A determined chance is not a chance, strictly speaking. This chance is not chaotic and random. The chance is more orderly than a lack of chance. The chance gives rise to a more deterministic world regarding purpose, meaning, and destiny. Destiny is the purpose of determination. But destiny, as all else discussed, is not necessarily determined. What is determined is that there should be existence, purpose, and meaning. From the point of view of purpose and meaning, the best possible existence is the existence responsible for its own becoming through chance.

The bigger question than determinism and free will or destiny is whether the chance is predetermined. As one of the fundamental forces of the Universe and its evolution (natural selection), the predetermined chance is not and cannot be proof against God's existence but precisely the opposite; regardless of how paradoxical this may sound, it is proof of its existence.

Chance does not act through some algorithm's random succession or order. No algorithm can be made or established of itself due to dead matter. If matter and energy of the universe were only matter and energy in a literal sense, they would be dead, completely chaotic, and incapable of any ordering and organization. But matter is not matter except as a manifestation of primary quality, as experienced by our senses. The only real essence and primary quality of the Being is something that philosophers throughout history sometimes called forms or noumenon, sometimes "aether" or what new scientists from our age like to call "the fifth force." Regardless of personal affinities and tastes, all these ideas must come up, and they usually do, with the concept of God in one way or another.

If chance is less accidental than we think and the world is less self-governed, the outcome is also the same for believers and non-believers. All arguments and counterarguments become worthless because of existence's vastness and paradoxical nature or the universe itself. Under such a scenario, all views or counterarguments would fit equally on either side.

What is natural selection? Do organisms develop due to an environment, or does the environment only trigger the potential to evolve in almost endless ways? What determines the survival of the fittest? How are the fittest organisms or animals formed? How can the first fittest animal be formed, and based on what? How can anybody, or anything, become stronger or better than anybody else or anything else, from the same material, under the same conditions? (Is fitness already there?)

AGNOSTICISM AND FAITH

Agnosticism is compatible with faith. We either believe or not, irrespective of our ability to find the truth. Faith, developed or established based on the nominal religion, is not proof of our ability to prove God's existence and is not proof of our implied agnosticism. Genuine faith is based on the belief in God regardless of the lack of evidence, religion, or agnosticism that may exist in a believer.

A believer is not necessarily someone who never had any doubts about the existence of God but rather someone who did not submit to skepticism and doubt. A believer is the one who fights agnosticism; an agnostic is the one who submits to it. To put it differently—a believer is the one whose atheism (or agnosticism) submits to his or her faith (or belief); an atheist is the one whose faith (if there is any) submits to atheism rather than submitting to faith or agnosticism; an agnostic is the one in whom faith and disbelief are equally present. This equidistance to God and atheism makes agnostics appear somewhat indifferent about these questions. Agnostics would rather wait for proof than bother tirelessly with the question without evidence. In believers, faith wins; in atheists, disbelief wins; in agnostics, neither belief nor disbelief wins.

An agnostic does not want to succumb to a religious heritage without proof supported by science. An agnostic is guided more by rational thinking than by emotions and religious sentiments that may or may not exist in him or her. This fight, or dilemma, in every one of us, is a fight between a believer and an atheist. Many believers are, perhaps, agnostics to some extent. They know they don't know the truth and will never find it, but they still believe (*credo quia absurdum*). Faith and agnosticism do not necessarily exclude each other, as it is usually perceived, and are compatible.

This confusion is caused by some, or many, agnostics refusing to believe or disbelieve God's existence rather because of their interpretation of agnosticism or God than because they are inherently suspicious. If proved, one way or the other, they would accept either way. They seek evidence for God's existence.

A believer rejects the idea of disbelief merely because it is impossible to prove God's existence or to know it in any other way except through faith. Although this may sound paradoxical, this approach is more rational than agnosticism because agnosticism excludes belief only because there is no evidence, even though providing that evidence is "impossible." On the other hand, nobody can say, prove, or exclude the possibility of such proof.

The biggest problem is not faith itself but the appropriation of faith by any particular religion or group, claiming to have the right (implicitly) to God. It is acquired through disputable knowledge and merely through "revelations" presented as truth. If only for this implicit claim, religions must be under constant scrutiny because such claims are not only logically false but are intellectually dishonest and immoral, regardless of the original intentions of their proponents. (Nonetheless, such ideas may have served an essential purpose, ethically and inspirationally, at the time of their making and for the particular people.)

All major philosophers are "right" to a greater or lesser extent. It's not merely a matter of right and wrong but of improving thought and discoveries that lead to a clearer understanding of the world. Every major philosopher provided a bit of understanding that clarified the crucial problems, even for those who objected to the ideas of the same philosophers.

Opponents of the particular ideas of past philosophers may be equally indebted, to a larger or lesser extent, to the very same philosophers they oppose as those they agree with. One thing is sure—there is no complete agreement or disagreement, nor can there be one, with any of the philosophers. (Complete agreement can be expected only within purely religious thought based on the dogmas of nominal religions.)

The main problem lies not in the beliefs but in the conception of God. What is God? Depending on the answer to the question of what God is, our relation, not only to God but to the idea of beliefs and religions, is formed and resolved. The main obstacle to this problem comes from our concept of God and not from God or the world itself. First of all, we cannot agree on what God is. We only see, analyze, and interpret religions in their expressed forms, primarily based on revelations that serve (and must serve) as God's given laws. In these books and "laws," God is described, ascribed, and prescribed. As such, God is a defined and untouchable being. The status of untouchability lies in revelations by the prophets, which is to say, in "God's own words." That is the only legitimacy to base these laws and secure them. We have no other fact or proof except the words of the few, which we must believe and follow.

What do we do by following these words or rules, laws, prophets and their revelations, and religions? We only believe in those who uttered and wrote these words and revelations. However, we do not know if there were any revelations in the first place or even if they were "real" to become the measures or expressions of truth or God automatically. There is nothing beyond this secret except our belief not in God but in the people (so-called prophets).

These practices persisted for thousands of years, and it is a logical consequence that our real idea about God, or our idea about "real God," necessarily became so distorted that we cannot talk about this idea or concept with almost any certainty. In this manner, we can never know if we talk about the same thing and mean the same thing unless we speak to people with the same opinions and way of thinking (followers).

Consequently, this implies the idea of many Gods (or many ideas of One God), and that's why there are many religions in practice and hundreds of religions in the past. The question, in religions, is not what God is, but who has the right to God (whose God is better).

CHAPTER 8

PHILOSOPHY, SCIENCE, AND RELIGION

There is a universal reservoir, the source of all knowledge, which is the same for the arts and sciences as for philosophy and religion. In a way, science is an art. In some ways, religion is philosophy because philosophy and religion often deal with the same questions—with the first and the final causes, among others. Religious people *a priori* "bet" on God, whereas philosophers may bet on God or not. The difference in approach toward God between religion and philosophy is that religion imposes and prescribes God, and philosophy offers the freedom of thought and choice. Religious prescriptions of God are not proof of faith or God's existence, but rather the opposite—they prove that philosophy is more "religious" than religion because it doesn't steal God from people but offers freedom.

Canonization, limitation, and reduction of God to a "few" sentences are inconceivable attempts to kill and expel God from people under the excuse that God is doing that. There is no weakness in science and language as instruments. There is no excuse. There is only human power and weakness. Understanding depends on the balance between human strengths and weaknesses.

The world was ready for greater discoveries two thousand years ago. (Heron of Alexandria[33] is an excellent example of an inventor who invented the first steam engine, *aeolipile*.) There was a basis

[33] Greek mathematician and inventor who lived in Alexandria, Egypt, in the 1st century AD.

for it. Only the language to translate abstract ideas and symbols into the language of science was missing. The only obstacle to a human being is a human being himself. The only limitation comes from the inability to dream and pierce into the essence that permeates all that exists in the universe. There is the same law for a galaxy, for a man, and for an ant. Basic principles are the same everywhere; they never change and exist as long as the universe exists.

God is not absolute, and Nothingness is not absolute. If we come from these positions (postulates), we confront the inability of language to be precise. This inability reveals that the problem of understanding different phenomena is not so much in the phenomena themselves but in our failure to understand or present them linguistically in the best and most precise manner.

God, or to put it better, our idea of God, is a concept before anything else. The way God had been (mainly) understood throughout history leads to the idea of God as an entity beyond the world that creates the world and stays beyond the world but affects it. There is no proof for this, but the concept itself, by its nature, is a theoretical expression or view of a particular phenomenon. As such, applied to something invisible or tangible and visible, it is subject to change.

Many concepts and ideas become laws in time, like the laws of physics. In time, many of these same concepts lose validity in favor of new concepts that become new laws. These phenomena lead to the progress of science; otherwise, some concepts, which were previously laws, would always stay the same regardless of whether they were correct.

At the same time, new concepts or laws would not be possible without the preceding concepts or laws. What we refuse today as incorrect is the basis for something we accept as correct tomorrow. Then the question arises—if the preceding concept was incorrect, how is it possible that the new concept, or law, that we think is correct can be based on the previous and rejected one? Something "incorrect" cannot be the basis for something correct. In this way, we conclude that even those laws (concepts) we reject today as incorrect contain the elements of a new law or concept or something that we think is correct or more correct. The world's universal nature cannot be wholly apprehended and observed by the human mind. All this happens while the laws stay the same; only our ideas and concepts about the laws change, not the laws or the truth (fact) itself.

Nothing in the world can be absolute except the Absolute itself. Everything else is contained and exists in relations. Relations can exist only through the division of the Absolute or God. Since no part is absolute, then no part can completely understand (including the human mind) or comprehend the whole of the Absolute. Understanding the "final" laws leads toward the end.
If humans understood and empowered themselves with the knowledge of these laws, they would cease to be humans and become what they sought to find or understand. In that way, humans become either God or nothing. The ultimate beauty is the evolution of the world. Evolution in the animal world (individual species) can be horizontal without major changes. Still, in the development of humans, it can be pretty vertical (thanks to awareness and brain power). Regardless of the vertical evolution of humans as biological beings, this is still only evolution, and evolution is only possible if something new is acquired, for the lack of final knowledge, about the final causes and the ultimate essence of being.

That which we never question is the concept itself. But what is a concept? Is it necessary to put the idea of a concept itself, or the law, under the magnifying glass to better understand it because a concept is only a word or term it expresses? Even the concept itself, as an idea or term, can be a matter of discussion or scientific, philosophical, artistic, or religious discourse. Consequently, we may discover that something that slowed down the development of ideas and their formulations or understanding of laws is the impossibility of breaking up with concepts or paradigms that became *a priori* truths.

In this sense, God is a concept more than anything else. The idea of God or about God, as it functions in religions, doesn't deal with reality but with presuppositions based on revelations. Revelation, in itself, is arbitrary and can result from hallucination, among other things, which does not exclude the possibility that even hallucination may lead to some ways of truth.

From the point of established religions, God is conceptually the source of the world, something that creates the world but stays beyond the world as something absolute that influences the world. Not only is God understood in this way, but it is also appropriated in the main monotheistic religions. Therefore, from the standpoint of official religions, it is much more important to claim God understood in this manner and establish the right to God than the God itself (accurate idea and concept of God). That is how this strange ownership functions and obtains "legitimacy."

> "Our drive toward unity obliges us to identify the world order of science with the God of religion. There is, however, this difference: for the religious man, God stands at the beginning; for the scientist, at the end of all thinking. We must believe to act ethically, and we must act; society could not survive if its members went about without proven moral precepts or waited until acquiring wisdom to decide how to behave. Therefore, each individual must strive to develop both sides of his nature, the religious and the scientific, which complete and complement one another: It is the steady, ongoing, never-slackening fight against skepticism and dogmatism, against unbelief and superstition, which religion and science wage together! The directing watchword in this struggle runs from the remotest past to the distant future: 'On to God!'"[34]

---

[34] Planck, Max. "Religion and natural science," in *Scientific Autobiography and Other Papers*, 187.

Divine Hiddenness Argument against God's Existence

Divine Hiddenness does not necessarily mean that the Ultimate Being hides; it instead means that our human powers are limited. The natural laws were secret (and still are) to the people before they learned to decipher them slowly. We may say that Newton's laws deciphered and formulated were hidden up to that point not because they were hiding *per se* but because our abilities were not on par with the laws of physics, which we have thought were "hidden." It is a poor argument to use hiddenness as a legitimate argument against God or against anything else, of which, at some point, we do not have a proper or complete understanding.

Hiddenness by itself is not proof that God does not exist. By using that logic, we may say that mosquitos are not aware of the existence of human beings. The argument that they are very "aware," in some sense while sucking the blood of humans would not be sufficient because they are not aware of who and what human beings are. Certainly, microorganisms, without any desire to compare human beings with microorganisms, are not aware of the existence of human beings.

What if animals used an argument, if they could, that there is no evidence that there are many galaxies in the Universe, or if any other animal could have used that argument? Would that be proof that other galaxies do not exist? On what basis are we sure that we possess the ability to experience God directly if it existed (although the world is one of the faces of God)?

I am not trying to compare human beings to other animals or diminish human abilities. Still, I would like to emphasize that, regardless of how advanced we are, we may still be as distant from God, or more, as some animals are from us. To rely only on evidence is to limit the science, not to be scientific. What is scientific in limiting science to the frame that fits our capacity for understanding, learning, and comprehension instead of fitting the frame of reality and the truth?

To be precise, we would need to redefine or make the idea of God more precise. Maybe God is not what we think it is. What if the World itself is God? What if the World, regardless of its beginning and end, is still a consequence of an eternal Being without a beginning and end? What if the world and matter as we know it are only the modes of the Universal Eternal Being from which everything originates and to which everything returns?

Matter is not what we think it is. God is not what religious books say. Nobody has the right to God, a title to God. No prophet can tell other people that he (or she) speaks the word of God. Humans do the things done in the name of God in their name, not the name of God. Their hiding behind God is a form of manipulation, demagogy, and control of others.

BERTRAND RUSSELL

In the following sections, there is no special reason for choosing Bertrand Russel and Stephen Hawking as two great representatives of philosophy and science, except that they are so symptomatic of both philosophy and science and knew how to express their ideas popularly and clearly. Although dear to me, I would not say they are necessarily my favorite philosopher and scientists, and the main reason for not exploring, to a larger extent, some other philosophers and scientists, closer to my heart, is the economy of space.

Russell's Teapot (Celestial Teapot Analogy)

We cannot equate Russell's teapot[35] idea with the idea of God.

[35] From Russell's unpublished 1952 article "Is there a God?": "If I were to suggest that between the Earth and Mars there is a china teapot revolving about the sun in an elliptical orbit, nobody would be able to disprove my assertion provided I were careful to add that the teapot is too small to be revealed even by our most powerful telescopes. But if I were to go on to say that, since my assertion cannot be disproved, it is an intolerable presumption on the part of human reason to doubt it, I should rightly be thought to be talking nonsense. If, however, the existence of such a teapot were affirmed in ancient books, taught as the sacred truth every Sunday, and instilled into the minds of children at school, hesitation to believe in its existence would become a mark of eccentricity and entitle the doubter to the attentions of the psychiatrist in an enlightened age or of the Inquisitor in an earlier time."

Although this idea is humorous, it isn't very sensible. If anybody without scientific credentials stated seriously that the teapot is circling the sun, the majority of people would think that a person stating that is either bipolar, schizophrenic or suffers from some other mental illness. This kind of comparison is absurd. Comic and absurdist comparisons of this kind only muddy the waters. Proof or disproof of such a thing is unnecessary because almost everybody knows the teapot can't orbit the sun as freely as planets on a microcosmic or macro level. Regardless of Russel being aware that his example is nonsense, he still used it (and he states that). The point was not to prove anything but to make a funny remark to diminish the subject of the attack, God. It is a logical fallacy whenever we use such tactics or tricks because we use witty comments for lacking something more potent. If we make fun of some ideas, it does not mean they have no value. We cannot destroy an idea that has existed for millennia by witty but silly arguments.

Carl Sagan made an even sillier argument about the undetectable dragon in his garage. To compare the idea of God to the teapot or a dragon in a garage is a useless way to refute an idea or argument with an "argument" (example) in the form of funny irony.

I must emphasize that I admire Bertrand Russel and Carl Sagan for their ingenuity and insights. I also admire Bertrand Russell's writing style because he could express complicated ideas and concepts in very readable and clear prose.

There can be no comparison between the idea of God and a teapot floating around the Sun or between God and an unidentifiable dragon in the garage. We cannot base our arguments on the value of their wittiness because regardless of how witty the statement is, it has to stand the test of truth, not the test of wittiness. We can easily exclude the idea of a teapot floating in orbit around the sun as ridiculous. The same applies to the argument about the dragon in a garage. But can we exclude the idea of God from religious and theological thoughts and serious philosophical inquiries interested in discovering the truth about the world and God? We can easily refuse to accept a teapot or dragon in the garage arguments as serious arguments. However, we cannot *a priori* deny the legitimacy of the idea about God, at least not the deist one (or pantheistic).

First, we must admit that we do not know everything about our world. What we know is not all that we can learn. We do not even know how much we do not know and how huge the difference between what we know and what we do not know is. We do not understand quantum mechanics in a real sense. We do not realize what gravitation is. We do not know if there is only one universe that we think originated from a Big Bang or if there are many other universes. We do not know what dark matter is, and we do not know what dark energy is, nor do we exactly understand these phenomena.

We still cannot say that what we don't understand about the universe or what we do not know about the universe does not exist. Our ignorance about the myriad of unknown and outstanding issues does not mean that these issues are teapots orbiting the sun or dragons or spirits hiding in our garages, apartments, or homes. There is only one home we all inhabit. That home is this universe, this world we live in. We learned much about our home but constantly pursued discoveries and new knowledge to bring us closer to the stars and the core of everything. That core of everything that we all crave hides beyond our cognitive abilities and shifts away again and again, regardless of our progress and discoveries. Are we closer to the truth today than Plato was almost 2,500 years ago?

Where is the core of the truth? If the center of the truth is out of space and time, then we cannot think about it from the paradigm of space and time. First, we have to define, as much as possible, the meaning of the word (term) God. What do we mean when we use the term God in every situation? We also must ask why God must be what some people say in the religious books. Why the Creator cannot be One with his creation? Why can't we assume that the World is God in action? The Source of everything is not only the Source but also the Materialization and Personification of itself through the world. Maker is the making and the creation of the making. Without making, the Maker is dead. Making and creating is the rejuvenation of the Maker. The world is another face of the Maker. The world is a way out. The World is God's exit strategy. The world is the salvation of the maker. Creation is the salvation of the Creator through its creation. Since the world itself is the source, the world is also uncreated. The world did not come out of nothing, *ex nihilo*. The world is one of the manifestations of the Absolute. The world is *proof* that the Absolute exists.

What we do not understand about the world or the things "beyond" the world is not proof that they do not exist but proof of our ignorance. Our ignorance cannot be a measure of anything and should not be a measure of anything, not to mention that it cannot be a measure of the unknown in this world or about what we think is beyond our world. Is something within or beyond only a matter of our understanding and perception rather than necessarily a matter of fact? The world is physics and metaphysics at the same time. Although it sounds paradoxical, we are what we are and are not. The world is more what we do not understand than we know. We seek what is seeking us. The Source is singing the song through all of us. The song of the universe celebrates the existence, the voyage, and the plurality dispersed through almost "eternity" to seek and reach the light that is the reflection of itself because the light of the source is in every one of us. To find the light within is to find the light out. Finding the source is identifying with the source on the way back to the source itself. The world is one organism. Existence needs separations to create and maintain life.

We would only treat Russel's view of the few main arguments for the existence of God here.

COSMOLOGICAL ARGUMENT

The First Cause Argument

The real problem with the first cause argument is not if God or anything else needed a cause but if the first cause is necessary for anything. If the Ultimate Being, or Absolute, always existed, nothing else exists or can exist outside of it except a transformation, "creation," or recreation within the Absolute. Absolute Being is not the first cause of the world. We must agree with Russel that the world is without a cause. The view that the Absolute (in someone's eyes, God) is without a cause does not mean that the world had a different, specific cause. The world is the Absolute itself or emanation of it. The world is the life of the Absolute (or God, but not a God from religious books), not something "different" from the Absolute. Therefore, neither the Absolute, God, Nothingness, nor the Universe need a cause because they are all the Absolute itself. Absolute is the causeless cause operating within itself. The world is a manifestation of the Absolute and its celebration, lovemaking between the Being and Nonbeing in the form of the cosmic fireworks for the hidden eyes of the Absolute. The World is a causeless cause's transformation, creation, or Recreation.

The Nothingness and the Being or Something are eternal and, therefore, are without a cause. The real question for atheists is how something can appear out of nothing at some point. Believing that something can come out from nothing is way less believable than the idea that there is an eternal Being that does not need a cause. We may call it whatever name we choose, but it is not necessarily incompatible with science. The limitations of science and its limited outreach cannot serve as proof against the eternal source of everything. Limitations are only proof of the level of science at some point. There is no absolute knowledge, and we can say, with almost absolute certainty, that humans cannot acquire absolute knowledge. Even if this were possible, humans would no longer be humans but be something else.

The Natural Law Argument

Bertrand Russell: "There is, as we all know, a law that if you throw dice, you will get double sixes only about once in thirty-six times, and we do not regard that as evidence that the fall of the dice is regulated by design."

Russell's argument is a logical fallacy because we cannot impose our understanding and interpretation of playing dice on God or the natural law. We must first define or understand our subject in order to talk about anything with scientific precision. Since nobody has an understanding of the world before the world, to put it that way, we cannot have a clear understanding or grasp of the things that are beyond our cognitive powers. We still can think about them. To say that science is only what is proven by scientific experiments would be foolish because that would exclude the vast space of the unknown, even unknowable. Maybe God does not play dice, but maybe even God needs, metaphorically speaking, to throw out thirty-six worlds to make some effects, even if only two, that would otherwise not be possible. As we know, matter cannot power itself and organize itself without the underlying creative force empowering it. Matter is matter thanks to our perceptive and cognitive powers, not *per se*. Matter *per se* does not exist in the form we see it. What we see is a reality based on our senses. We cannot completely rely on our senses to tell the underlying reality. Reaching the underlying reality is possible only through abstract thought. This abstract thought will enhance scientific discoveries because we cannot reach the physically unreachable by experiments or strictly scientific means.

Identification of God from religious books with God independent of holy books is prevalent in the books or arguments against God used by the most famous atheists, including agnostics like Bertrand Russell.[36] However, a huge difference exists between a God from religious books and Spinoza's God or the God of many philosophers and scientists. Once we acknowledge and accept this important difference, we will realize that the gap between believers (not contaminated by religions) and atheists (or agnostics) is much smaller than it looks at first sight. God is not in the religious books, nor can he be owned through religious books. The main goal of the major monotheistic religions is to *a priori* appropriate and establish the right to God rather than to define and explain God in the deepest possible sense because that is almost impossible, even for science and philosophy. For that reason, a belief in blind faith and fear mostly saves major religions, rather than pure belief, unaffected by religious influence or deceit.

The Argument from Design

Based on Russell's treatment of this argument, we assume that Russel expected that the world's creation, by design, had to be perfect. But, as with all other arguments, we must establish what

---

[36] Russel, Bertrand. *Why I Am Not a Christian*. 1927.
"If you say, as more orthodox theologians do, that in all the laws which God issues he had a reason for giving those laws rather than others -- the reason, of course, being to create the best universe, although you would never think it to look at it -- if there was a reason for the laws which God gave, then God himself was subject to law, and therefore you do not get any advantage by introducing God as an intermediary. You really have a law outside and anterior to the divine edicts, and God does not serve your purpose, because he is not the ultimate law-giver. In short, this whole argument from natural law no longer has anything like the strength that it used to have."

design and perfection mean. If we do not clearly define what design is and what perfection is, we are applying our judgments to something either undefined or loosely defined. Evolutionary theory, be it Darwin's theory, cannot be proof of a bad design of the world. Anomalies or shortages in the world are not proof of a bad design. Imperfections are needed in the world and serve a higher purpose. Let's say that God, if he existed, wanted to create the perfect world. This perfect world would be sterile. In the perfect world, there would be no cosmic hierarchies, lows, and highs, enough friction to sustain life as something whose purpose is not to be made perfect from the beginning but to seek perfection, to make "progress" in myriad ways toward the main purpose which is life itself. Life, by definition, is not perfect. Perfect life is not a real life.

The purpose of design is not to predict a Ku Klux Klan or the fascists and eliminate them from the design before any creation but to put the "engine" of the vast Universe in motion, to enable the world to seek its paths freely, without a God playing dice. That is where determinism and free will come together to create a sensible world.

Design does not mean playing dice, nor necessarily creating something new, but the creator offers himself an exit to exist in an ever-new world, a new form with meaning. We also may say that in the Universe or Omniverse, beyond our knowledge, there can be not only thirty-six (to make a comparison with dice) but a googolplex of universes (dice), and the possibility for combinations is infinite.

"Impossibility to prove God" is not proof that God does not exist. Russel would argue that the burden of proof is on the person making a claim, but the world itself is proof of God's existence. The solution to this enigma is to recognize that the world is God. The problem is not belief or disbelief, first cause, natural law or good or bad design, or any other argument for the existence or against the existence of God; the problem is in our understanding and consensus about the idea of what God is. Argumentation or proof can never be shifted to only one side. Something so obvious as the world does not need proof but understanding that the world is also, in its deepest nature, God itself.

We can fight as long as we want, but if we fight from different positions for the sake of different positions, we are not going anywhere. God is not the same for the theist or the deist. Christian God is so far from Spinoza's idea about God. The majority of people who are atheists today are atheists more in revolt against nominal, official religions and not necessarily in revolt against God if this God was better defined or approached from an angle unaffected by religions.

STEPHEN HAWKING

We will start with Hawking's few quotations.

> "The quantum theory of gravity has opened up a new possibility, in which there would be no boundary to spacetime and so there would be no need to specify the behavior at the boundary. There would be no singularities at which the laws of science broke down and no edge of spacetime at which one would have to appeal to God or some new law to set the boundary conditions for spacetime. One could say: 'The boundary condition of the universe is that it has no boundary.' The universe would be completely self-contained and not affected by anything outside itself. It would neither be created nor destroyed. It would just BE."[37]

Or, in the same manner: "There ought to be something very special about the boundary conditions of the universe, and what can be more special than the condition that there is no boundary?"[38] Also, he stated, "According to the no-boundary proposal, asking what came before the Big Bang is meaningless—like asking what is south of the South Pole—because there is no notion of time

[37] Hawking, S. (1988). *A Brief History of Time* (p. 136). New York: Bantam Books.

[38] Hawking, S. (1981) Speech on the origin of the Universe in the Vatican.

available to refer to. The concept of time only exists within our universe."[39]

The "no-boundary proposal" is a classic example of a device called in Latin, *Deus ex machina*—God from the machine, invented by the ancient Greek dramatists Aeschylus and Euripides. The primary purpose of the device was to resolve the irresolvable.

The question of what came before the Big Bang is not meaningless. We cannot accept that our Big Bang is the beginning of all existence. Since there is "no notion of time available to refer to," that does not mean there is nothing to refer to. This reasoning is a logical fallacy based on the idea that there should be nothing to refer to if there is no time to refer to it. This kind of reasoning falsifies reality to fit the argument. For this statement to be accurate, there must be proof that there is nothing to refer to, not "no notion of time to refer to." The lack of notion of time to refer to or its availability is not proof that there is nothing to refer to, but only that there is no notion of time to refer to and that it is not available. The lack of availability is only proof that something is not available to someone but not proof that nothing exists beyond the "point" where "time" stops. If Something, the Being, the Universal Source of Everything, is not available or approachable in any way by some particular scientist, that does not mean that the Universal Source of Everything (the Absolute) does not exist beyond the physical world.

In this sense, the no-boundary proposal is a boundary proposal of a different kind. Since it is impossible to speculate about abstract concepts or ideas, such as God, Absolute, or Universal Source, it is easier to invent some trick (pardon my language), with all due respect, to compensate for the lack of understanding of the

---

[39] Hawking, S. (2018, October 16). *Brief Answers to The Big Questions*. Chapter 2: "How Did it all Begin?" Last paragraph: "What came before the Big Bang?" New York: Bantam Books.

most abstract ideas and to compensate for the limitations of a frame of mind of any particular scientist or philosopher. In this case, the no-boundary proposal precisely serves the purpose of a boundary—to limit the world to the point where "time stops" and declare that there is nothing beyond because time stops there. That should mean that the laws of nature and science stop at this artificially produced boundary. But what do we have as proof that this is true? Precisely like in religions, we have words that sound seductively beautiful and convincing. Also, to a large extent, these words are supported by scientific knowledge and investigation. Yet, they are just words, and in no way do they prove that there is no immaterial Universal Source beyond the "point" where time stops.

Hawking's insistence on scientific laws hides the desire to transform the current scientific laws into the ultimate and absolute knowledge of everything, physical and metaphysical, of this world and the outer world. According to him, we are on the verge of declaring, with almost absolute certainty, that we have solved the whole enigma of existence and gone down to nearly the deepest end of science and scientific laws. Although he expressed many ideas in a simple, popular, and often funny way, there is a little bit of unjustifiable scientific conceit (to call it that way) behind some statements.

If we were to imagine the creative force capable of creating the Universe, this creative force would be out of time or eternal. The Eternal Being is not contingent or affected by the boundaries of the physical world. The no-boundary proposal is accurate in that there are no boundaries we can apply to the Eternal Being. Still, the Universe, as the Being with its beginning, is bounded by time. The first point of the Universe is its first limit; it would have no limits if it were a timeless Being. Even if there were a series of births and rebirths, these would still be limited creations or recreations of something eternal that creates or recreates itself through the creation of universes. The creative power of the Eternal Being is the ultimate force that keeps the Eternal Being alive. The only way for the Eternal Being to exist with meaning is through its creative power to rejuvenate itself in new ways and myriad forms constantly. The Creator is its creation, and the creation creates the creator in a deeper sense. Without creating, the Universal Being loses its purpose and becomes meaningless. Meaning is only possible in plurality. The World, or Universe, gives the Universal Being meaning and purpose. The world is its salvation.

Everything or anything before the Big Bang does not obey the laws of physics and is not necessarily compatible with our understanding and views of the world. We do not mean that there is nothing before the Big Bang, but that we can hardly understand what is hidden before it (we can only use our abstract thoughts to imagine it). As we explained in other parts of this book, we cannot look at time in the same way from the perspective of the "physical" world as from the perspective of an immaterial world. Neither space nor time exists in the nonmaterial world. Space and time are categories we deal with only in the "physical" world.

The nonmaterial world is the spaceless "space" and timeless "time." In the nonmaterial world, there is no time and no space. Paradoxically, no time means eternity, and no space means infinity. No boundary proposal is a proposal about an everlasting world and everlasting time. Everlasting means time with a beginning and no end. The concept or idea of eternal implies no beginning and no end. The concept of the everlasting Being, or time, or space, as an ultimate principle is shaky because we cannot use our limitations in thinking to argue against reality as it is and not as we say it is. If we do not understand the idea of time in its totality, we cannot talk about time from the point of superior "knowledge."

Whatever has a beginning is not eternal. What is not eternal must have come from "somewhere." The worst thing is to try to prove something, counting on the idea that it came from nothing. It is also as unscientific as it can get. No chaos theory can prove this reasoning, and no "butterfly effect" can prove it because nothing is only nothing and cannot be anything else except nothing. Nothingness cannot move the Nothing or create from nothing.

The idea of infinite regress is absurd. The question of God cannot be reduced to a cosmological argument either. The ultimate question of reality is if there is something or not. If we agree that there is something, then the question is if that something can come into existence from nothing. Wouldn't it be more logical that there is just nothing? But just the idea that there is nothing implies, at least linguistically, that nothing is something; otherwise, we would not use the words ‘there is.’ Again, language demonstrates how limited it is.

We should conclude that this Nothing is eternal. To prove that Something is not eternal, we would have to prove that this Something just appeared from nowhere and became something. The burden of proof here lies on atheists to demonstrate how something came into existence from nothing. This burden is the same one Bertrand Russell tried to impose on theists, deists, or any believer. Why would believers be burdened with the burden of proving God and disbelievers get away without proving how Something, whatever we choose to call it—God, Universe, World, came to be?

That the matter originated just by itself or was always "there" in the form of "energy" and organized itself into superbly sophisticated organisms forming the Universe borders on insanity. We can, perhaps, all (both atheists and those who believe in a higher source) agree that there always was Something. The question is how we define this Something. It is much easier to prove that the Something always existed than to prove the opposite. Scientists can deal with this Something more easily because, moving back, step by step, scientists will come to nothing. When they come to nothing, the "spotless spot" before the Big Bang, they will have to scientifically explain how all the reality, the whole of what we think the "Universe" is, was contained within an immeasurable "spotless spot." Once they find the answer, they will understand that this "spotless spot" from which everything originated is immaterial and spaceless. This immaterial Being is the Creator of all reality and is the reality itself. Plotinus would call this reality intelligence or mind.

We provided the frame from which scientists can prove that matter is a construct, program, and "instruction" of the Universal Mind and that the same program predetermines our perception. Our understanding of the world is contingent upon our experience, cognition, and perception (tertiary quality in my system of thought), which is contingent upon the secondary in my system of thought (originally, primary quality) since there is no matter as we perceive it or conceptualize it. There is no matter as such.

Regarding this, religious thinking is more accurate in realizing there can be no creation without the mover. Zero point is the "first" point. This zero point is something. It did not come from nothing. The World, as a "physical reality," has its physical and temporal beginning, but the force empowering it, the Universal Mind, is out of time, so there can be no notion of time to refer to before the Big Bang.

The comment about the South of South Pole is a logical fallacy because it presupposes nothing existed before the Big Bang. The Entity before the Big Bang is not comparable to what is south of the South Pole. The idea that we do not understand that there is something "North, South, East, and West" of the Big Bang does not mean that these "North, South, East, and West" do not exist in a way incomprehensible to our limited cognitive powers, senses, and based on limited scientific discoveries. That which lies beyond time and space and is the source of everything we see cannot be referenced only by the too-simplistic and sometimes semi-humorous means. Our language and words or terms describing the sides of the "physical world" are limited if we want to apply them to the metaphysical, immaterial realm beyond space and time. In no way does this kind of reasoning or "arguments" prove that the

Universal Source of Everything does not exist. I have chosen to call this Source of Everything, known and unknown, the Universal Mind. In our sense of the word, there is no space before the Big Bang, and there are no sides of any kind, not to mention sides of the World. South of the South Pole is just Nothingness from a physical point of view. There are no sides of the world in the primordial Nothingness or the Absolute Vacuum. At the same time, this Nothingness "hides" immaterial Being, the Universal Source of everything.

No Need for Time Before the Creation of the World

Here is one more quotation representative of the way Stephen Hawking thought:

> "The role played by time at the beginning of the Universe is, I believe, the final key to removing the need for a grand designer and revealing how the Universe created itself. As we travel back in time towards the moment of the Big Bang, the Universe gets smaller and smaller and smaller, until it finally comes to a point where the whole Universe is a space so small that it is in effect a single infinitesimally small, infinitesimally dense black hole. And just as with modern-day black holes, floating around in space, the laws of nature dictate something quite extraordinary. They tell us that here too time itself must come to a stop. You can't get to a time before the Big Bang because there was no time before the Big Bang. We have finally found something that doesn't have a cause, because there was no time for a cause to exist in. For me this means that there is no possibility of a creator, because there is no time for a creator to have existed in. People want answers to the big questions, like why we are here. They don't expect the answers to be easy, so they are prepared to struggle a bit. When people ask me

> if a God created the Universe, I tell them that the question itself makes no sense. Time didn't exist before the Big Bang so there is no time for God to make the Universe in. It's like asking for directions to the edge of the Earth—the Earth is a sphere that doesn't have an edge, so looking for it is a futile exercise."[40]

In its absolute state, beyond the World, the Being is immaterial, and the Nonbeing is an absolute vacuum, nothingness, or emptiness.

In the primordial state of the Absolute, the Being and the Nonbing become the same—the Nonbeing.

There is no time or space in the absolute realm beyond the World.

Timeless "time" is the potential for Eternity. Eternity is beyond time because it is all time, past and future.

Spaceless "space" or nothingness is the infinite potential for space. Infinity is beyond space because infinity is all space, past and future.

Creation or recreation of the World (Universe) activates the two poles of the Absolute.

Creation of the World is the salvation of the Absolute.

Absolute is absolute potential.

The activity of the Being enveloping the Nonbeing (Nothingness) transforms the Being and the Nonbeing into the World (Universe).

---

[40] Hawking, S. (2018, October 16). *Brief Answers to The Big Questions*. Chapter 1: "Is There a God?" New York: Bantam Books.

When the Absolute transforms into the World, the Being becomes positive, and the Nonbeing becomes negative.

The Being is positive "energy."

The Nonbeing is negative "energy."

Zero is the point of equilibrium between the Being and the Nonbeing.

Zero is the passage (wormhole) between the primordial state of the Absolute and the World or Universe.

Before the spacetime continuum, plus and minus are the same: + = –

Before the creation, Absolute is **0 (+ – = 0)**

At the Creation of the World, the Being envelopes the Nonbeing: + **0** –

The Being cannot envelop the whole of the Nonbeing because Nothingness is infinite in its potential.

The Being is infinite in its potential, too.

(+ [plus] is the Being; – [minus] the Nonbeing; **0** [Zero] is the Absolute)

The primordial state of the Absolute is immaterial, spaceless, and timeless.

The primordial state of the Absolute is absolute potential.

In its potential, the Absolute is infinite and eternal.

Absolute can only exercise its potential and power in the infinite number of possibilities and universes or worlds it can transform into.

"Point" of creation, like a Big Bang, is the passage through Zero between the immaterial, "timeless," and "spaceless" Absolute into the Universe with space and time and with characteristics of physical reality, which is an illusion. But this illusion is reality, not less realistic for being an illusion.

In this sense, there can be no time before the Big Bang, but time and space are logical consequences of our Universe appearing after the Big Bang (or a different event) or any other universe.

The power of "negative energy" is the power of the Nonbeing or Nothingness, and the power of "positive energy" is the power of the Being.

The Universe is not material, as we already described in the previous chapters of this book. Hence, the annihilation, real or unreal, is irrelevant because when positive and negative "energy," if that is the case, cancel each other out, it is not the annihilation of the Universal Source, the Absolute, or its two constituents—the Being and the Nobeing (nothingness). Everything is a program. Cancelling out of the positive and negative "energy" is annihilating the original immaterial program. Simply put, only the "illusion" of the physical world disappears.

Nothing can ever come from nothing.

Universal Source and universal potential are always beyond space and time and do not abide by human scientific laws. Still, this does not mean that, potentially, humans may not acquire abilities to solve the biggest mysteries within the boundaries of the existing Universe we are aware of or of the much larger Universe this one is only part of. Even the secrets hidden beyond the "point" where time stops may be potentially solved, at least by abstract thinking, which can be a vast and valuable source for the further development and enrichment of scientific laws and humans as a species.

There is no need for time before creation because the Universe is not created in advance and thrown into the abyss of nothingness. No, the World's creation starts precisely with space, which is also the creation of time. Instead of stating that there is no need for the creator when time stops, we say that before the creation, there was no need for time because the World had no "physical" appearance then. The Universe is the physical appearance in the spacetime continuum of its metaphysical, immaterial essence beyond the spacetime continuum. This "physical" appearance is only an illusion. However, this illusion is even more real than if it were really and strictly physical without any programming and interconnectedness between the three kinds of qualities or modes making up the reality we experience.

Reinvented and Redefined God

What are scientific laws, or natural laws for that matter?

It is wrong to make statements in the name of science to presuppose the truth instead of accurately representing it.

It is wrong to equate scientific laws, as a human construct, with the absolute truth.

It is wrong to use statements such as "there is no south of the South Pole" as proof that there is nothing beyond the point when time stops. Such statements may sound seductive, but they are not scientific, nor do they prove what they try to prove.

Similarly, it is wrong to apply the laws and states of our Universe to the state before the Big Bang if we do not have the slightest idea about the "context" and the state before the Big Bang.

The lack of time and space in our sense of these concepts does not mean that nothing exists beyond the spacetime continuum.

To state there is no need for a creator or God if there is no time is meaningless. Because the Absolute is beyond time and space as we know them. Absolute is the Source of time and space and, therefore, does not need time to create or recreate itself through the World.

Such statements are not scientific; they are presuppositions and intellectual constructs that serve as fillers for the lack of accurate understanding or a fundamental theory explaining how reality, which is almost incomprehensible to the human mind, functions.

There are laws of nature. We may describe the laws of nature but still not wholly unravel and understand them to pronounce that scientific laws are equal to natural laws or that they are an absolute representation of natural laws. Scientific laws did not yet cover, describe, and explain all natural laws, mysteries, and secrets.

We cannot hide behind science nor talk in the name of science while proclaiming statements that are not scientific. Such statements are that there is no creator if there is no time. However, this is only a statement a scientist (Stephen Hawking) uses, not a scientific argument or fact with scientific validity. The same goes for the south of the South Pole because such a statement is a logical fallacy based on the presupposition that there is nothing south of the South Pole, meaning there is absolutely nothing beyond some point of the material World, which is only a presupposition and not a scientific or natural law. Such statements may sound like analytic or logical propositions. Still, these are not analytic or logical propositions but methods or tricks to cover up the loopholes of scientific knowledge and understanding with dogmatic statements.

Science must be antidogmatic. Only the church can be dogmatic. It would be better for the church to become more scientific and less dogmatic to better understand what God is before representing something created by man, not by God, in words written by man, not by God.

Science is a human construct. Scientific laws are as valid as our understanding at any particular moment. Scientific laws are not absolute. Scientific laws wear human colors. Regardless of how close we got to the truth, the scientific laws are still not 100% accurate and complete. If we do not possess absolute truth, then scientific laws cannot be final, or they may be representative of the truth up to the level of our knowledge and understanding, not more.

We may accept the idea that natural laws govern everything in the Universe and that these laws are in some way absolute. Still, we cannot hide behind scientific laws before explaining them. According to Hawking, scientific laws may be enough for our understanding of the World. His implicit message is that the Creator is not needed. Such statements could have been valid if scientific laws were absolute and scientists, including Hawking, resolved the mystery of existence, the Universe, and the origin and future of everything. Since that is not the case, no scientist can replace the idea of the Creator just by insufficient scientific knowledge. Only a scientist or scientists (or anybody) with absolute knowledge can dethrone the Creator if there is such complete knowledge (scientific or otherwise), proving that there is nothing beyond the "point" where time stops. Unfortunately, this kind of knowledge and understanding does not yet exist. The purpose of science is not to push the Creator out of the picture but to improve, define, and redefine scientific laws in its pursuit of truth.

One of the tasks of science is to redefine the word and concept of God so that we can come up with an approximate idea about what God is, at least in scientific and academic communities. For instance, the Absolute (Universal Mind) is the Source of Everything, and it exists regardless of whether time stops at some "point" and irrespective of the no-boundary proposal. The no-boundary proposal is one of the biggest boundaries to the science ever produced in history because it tries to impose the limitations of a particular human mind onto the whole Universe and beyond and declare this limitation, boundary, to be a scientific law.

Also, we must redefine and come to a consensus on what God (Creator) is and what creation and recreating is. When we do this, we may make the gap between believers and atheists smaller, and we may realize that faith and atheism are compatible. The major incompatibility comes from the preprogrammed thinking and concepts based on religious books. Most of these concepts humanity accepted without real consideration and analysis, betting on the idea of a God that I call the stolen God. A stolen God is not a “real” God and cannot be used against the “real” God if it exists. The idea of a stolen God gains most of its power from the fear of millions or billions of human beings belonging to the major monotheistic religions.

Based on my theory, the Absolute, Supreme Being consists of the Being and the Nonbeing. The Being is part of the Absolute and is not absolute, but the Being is that which was traditionally considered God. The creation of the World is almost equally contingent upon the Being as it is upon the Nonbeing (nothingness). Not only can God not be omnipotent without the Nonbeing or absolute void, but the World’s creation depends almost equally upon these two poles of the Absolute. The Being is the positive pole of the Absolute, and the Nonbeing is the negative. Zero **(0)** is the wormhole between the Absolute immaterial realm of reality and material reality or the Universe. The *Zero*, as such, is the Source of Potential Infinity, the *Perpetual Motion Machine of Existence*, and, in a way, the Absolute itself.

The Creation of the World or Omniverse is the Creation or Recreation of the Absolute itself through the two poles, two “sides” of its Self, the Being and the Nonbeing. Without the creating, everything is the Nonbeing. Creating is the Savior of the

Absolute and the Being. Otherwise, everything transforms into absolute Nothingness or Nonbeing. Since creating is the recreation of the Creator, it is the activation of the absolute potential to recreate itself in a new way.

No time or space exists beyond the Universe and "physical" World.

The spacetime continuum is the worldly reality.

Beyond the spacetime continuum is the metaphysical Reality of the Absolute (the Being and Nonbeing).

Zero is Absolute

+ = **B** (the Being or God)

– = **N** (the Nonbeing or Nothingness)

+ – **(Being-Nonbeing)** = **0** (the Absolute)

At the moment of creation, the Being envelopes the Nonbeing and transforms into the World through the wormhole of Zero. That is the creation of the World or "partial" transformation of the metaphysical into "physical," although the physical is only an appearance (illusion).

Zero is Absolute Potential

In the metaphysical state:

Zero is the present.

Zero is the representation of the infinite and the eternal.

Zero is the passage from the metaphysical to the "physical" realm at the moment of creation. At this point, the infinite and eternal transform into the categories of the spacetime world.

Zero is also a passage from the past to the future. At this point, the present stops because the present is eternal. The present is only a passage from the past to the future and is only possible in spacetime.

How we see and perceive past and future, or time in a physical sense, exists only in the physical World. Beyond this World is a metaphysical World where there is no spacetime continuum. At this point, time stops.

When the spacetime continuum stops or disappears, past and future disappear, and all past and future are in an absolute Zero as Absolute Potential. All time and space are in Absolute Zero, which is infinite and eternal. Only this Absolute Zero beyond the spacetime continuum is absolute time, which is present, and absolute space, which is nothingness.

Zero in the metaphysical realm is Potential. We may call it, conditionally, passive Zero.

Zero in the spacetime realm is active Zero. In this realm, the present is only a passage from the past to the future. When time stops, it becomes absolute present, which is absolute time or eternity. The Source of all time and space is beyond time and space.

LANGUAGE AND PHILOSOPHY

Language is the most efficient instrument with which humans articulate and communicate thoughts. This statement is generally true, but if unconditionally accepted, it could lead to a misunderstanding of the world of thoughts, ideas, emotions, and visual experiences, ultimately leading to an insufficient understanding of the world.

Philosophy may be sometimes hard to grasp, but it is profoundly meaningful. Mudding the waters under the pretense of doing something meaningful and great is meaningless. Language is one of the best tools of expression, but also one of the best tools to deceive, manipulate, and lead others astray. It is sometimes hard to differentiate between the two and recognize the game of deceit and manipulation. It usually happens when we need to figure out our real prowess and use the language not for real communication or expression but for personal gain, academically, or in any other way that may benefit us.

Humans free themselves from conceptual traps by freeing themselves from paradigms that close horizons instead of opening them. Language is the supreme instrument of thought, but a number is a word, too; without a word, there would be no number. Every number corresponds not only to the graphic symbol but also to the linguistic one. Conceptually, numbers and words are different because words represent or name things and phenomena, while numbers represent the quantitative or numeric value of things and phenomena. We understand the function and the role of numbers and words. Through words, others know what we think or want to say. However, even a bird knows what the nest is by feeling it intuitively. A dog, thrown away ten miles from the house where it lived, will find it. Animals often communicate among themselves.

A word had to predate the number. As a graphic sign, the number is the graphic expression of the word that predates it. Both words and numbers are linguistic expressions of thought. The purpose of words is to express and explain the essence, and the purpose of numbers is to express quantity. Words are qualificators, and numbers are quantifications.

Considering all this, we must reaffirm the great potential of language as an instrument of communication and for expressing the most complex thoughts and ideas. Not only that but in this way, language affects and shapes our thinking to a degree. At the same time, we must reiterate the limitations of language regardless of our desire to be as precise as possible. We can be sure that language is relative. Since language, as an instrument, is not absolute, there can be no absolute way of defining things. In this sense, we must learn to accept implications and hidden meanings as values equally crucial as the obvious ones. Hinting at something is more important and valuable than volumes of sterile compilations of strange words without a more profound sense and meaning.

We may also think that, regardless of how insufficient our language or knowledge is, we can still arrive at conclusions stemming from something more profound than the knowledge or language but only expressed by language. If language (as we understand it) were an absolute instrument of thought, there would be no different languages but only one universal language of thought. Since this is not the case, we can express the same thought in different languages with different signs and sounds.

Let's agree that spoken language is not an absolute instrument of thought. Still, we must agree that language, as such, serves the profound purpose of communication primarily (on a fundamental level) and some other, more sublime purposes (great works of literature and philosophy, for instance). The primary purpose of language, as one of the phenomena of existence, must be analyzed as a phenomenon with its existence and essence. Only when defined with the highest degree of accuracy, can its essence be applied to existence as the totality and the world.

(It is possible to imagine nonverbal communication of thought on a much higher level than the one we possess. We have to determine first what the shortcomings and advantages of verbal communication [and communication in general] are.)

Although language, as we know it, helped humans acquire supremacy over other animals, if we follow evolution, we could easily infer that language is still only a purely animalistic expression, regardless of how advanced it seems to us compared to other animals. But what the real potential and possibilities are, we barely know or try to explore and understand.

It is a different and huge subject if language or communication must be in the way we, as human beings, think language is, how it functions and expresses thoughts. Different beings with intelligence, somewhere in the Universe, use different types of communication than our own, not only in their languages but also in how they communicate. Depending on their stage in development, their communication abilities would be more advanced. We can only imagine the behavior or communication among the beings belonging to a civilization a million years or more ahead of our own. We can imagine that if we, as human beings, were able to survive that long, in the distant future, we would be so advanced compared to us now that it would be like comparing different species.

Can thought exist without words? Are visions, for example, visual thoughts? What would, potentially, be the most sublime form of language and communication? Is thought possible without language in our sense of the word? What would be a pure thought?

There is, likely, a potential far exceeding our current abilities and ideas about language. If higher beings in the Universe possessed these powers and abilities, they would be close to those described in the *Bible* as angels. Indeed, beings so advanced would appear to us as higher or supreme beings, perhaps transcending physical existence in a way. In that sense, their communication may be so sophisticated that the appearance would be as if pure thought communicated with other pure thoughts.

APPENDIX
SOME SCIENTIFIC FACTS AND COMMENTS

The last section of this book contains data and information that may help provide a better understanding of the book. I realize this information is unnecessary for philosophers or scientists working on these topics. Still, it may benefit an ordinary reader who is not necessarily familiar with these ideas and concepts. Since this information, references, definitions, and explanations are not mine but are generally accepted and widely known in certain academic circles, I did not consider it essential to emphasize this in the form of footnotes. For this reason, whenever I have used such information, explanations, or "definitions," I have italicized them instead of using footnotes. I have not used italics in my comments.

- *The string is the smallest thing in the Universe.*
- *Plank length is 20 factors of 10 smaller than an electron* (10^{-33}).
- *Can space and time exist at the Plank length? Scientists do not believe that this is possible.*

Comment: If our entire Universe is smaller than the Planck length before the Big Bang, that is proof enough for me that it is immaterial and powered by the original immaterial Universal Source (the Universal Spirit). This is the point of appearance and disappearance.

- *According to Gabriele Veneziano[41] (1942), the Big Bang (13.7 billion years ago) was not the beginning of the Universe. He believes there was something 10 to 120 bigger than the Universe before the Big Bang.*

Comment: This thing that is 10 to 120 times bigger than the matter in the Universe we know is the Universe I am talking about. It could even be much larger than 120 times larger than the matter in the Universe because the effects we can measure as dark matter and dark energy at a point infinitely far away may not be measurable or at least not adequately measurable.

Black Holes

- *Gravity is infinite at the center of a black hole "where" space and time do not exist.*
- *According to Einstein, a singularity is a point of zero size and absolute density in the heart of a black hole. It has no space at all.* (I agree with this.)
- *Edge of a Black Hole. In the fifties, physicist Wolfgang Rindler (1924 – 2019) invented the term event horizon, a boundary beyond which events cannot affect an observer.*
- *The gateway to a black hole is a wormhole.*[42]
- *Inside the black hole is an unstable wormhole inside a deadly spinning ring.*
- *Einstein said that if you fall into a black hole, a gateway connects you to another universe, but he didn't believe it would be possible to break through.*
- *According to Gerardus (Gerard) 't Hooft[43] (1946), information from the asteroid will be saved on the surface of a black hole. All information is on the surface.*

[41] The father of string theory.

[42] John Archibald Wheeler (1911 – 2008), American theoretical physicist, coined the word wormhole in 1957.

[43] Shared the 1999 Nobel Prize in Physics.

- *According to quantum physics and the conservation of information theory, information cannot be created or destroyed. Nothing that exists can be obliterated. On the other hand, there is a black hole paradox: when a black hole evaporates, information completely disappears.*

Comment: Based on my idea of the Universal immaterial Source of everything, the basis of all information and all existence can never disappear as an ultimate potential. Nevertheless, we cannot be sure that the information about all particular existences and Universes is saved. The question is more if there is a reason for that than if it is possible. For example, if the whole universe, like ours, disappears, would it be necessary that all information is saved? The sole existence of the universe is information. The destruction of "matter" does not mean the original information on which "matter" functions can disappear. What disappears is the illusion in the form of energy and matter. But primordial information, or the Source that provided information for forming energy and matter, is not the memory of what is created and cannot be lost. I believe what is used initially is saved, but the universe's memory is lost. Still, if we look at it from the point of absolute potential, this memory is "not lost" because it can be "recovered" or repeated in the working of the absolute potential as a recurrence. Absolute potential means that the same universe with the same "life" and evolution can exist again.

- *Negative matter causes anti-gravitational force and a negative gravitational field.*
- *Equations don't work at the level of quantum physics.*
- *Einstein's theory of gravity works with the big and quantum with the small.*
- *Quantum theory and relativity are not compatible.*
- *Quantum gravity is a quantum version of the theory of relativity.*

- *At the speed of light, the mass becomes infinite while the rest mass stays the same.*

Comment: When, at the speed of light, the mass becomes "infinite," that is the moment of absolute, infinite density when space disappears. "Infinite" density is the "point" of disappearance. The spaceless point of zero is the channel between the Being and Nonbeing, between existence and nothingness. Through this zero point, the Universal Mind "materializes" itself by creating "matter" and the existing universe. That is the "point" before the Big Bang, the point of absolute density and no space or time. Absolute density is the "point" before the dispersion of Oneness (Singularity) into materialized plurality.

Infinite mass is impossible. The state of absolute mass would be when everything would transform into mass, and the void would disappear. That is impossible. Reaching the state of infinite mass is the same state the "mass" or "energy" was in before the Big Bang. That state is the state of no mass and no energy. Absolute "density" is the state beyond matter and energy. This state is immaterial. The effect of the infinite kinetic energy would be equivalent to the infinite mass if possible.

- *When the body is outside the gravitational pull, its kinetic and potential energy are zero. According to Newton, gravitational potential and kinetic energy become zero at the point of* $r = \infty$.

Comment: This gravitational pull is, actually, the Max Plank's vibration, which, as a source of motion, is the source of the gravitational pull, without which kinetic and potential energy would be zero. At this point, everything stops. At the speed of light, an object has infinite kinetic energy, which equals infinite mass. This infinite kinetic energy or infinite mass is static and massless. That is the point of absolute density. The world becomes static when its mass reaches the point of absolute speed, which would be equivalent to the same infinite point reached by the mass traveling at the speed of light. This is not infinite mass but an effect of the kinetic energy produced by the speed of light equal to the infinite mass. This proves my point that everything would stop if it were not for the kinetic energy and the "gravitational" pull fed by the immaterial Universal Source of everything.

- *Kinetic energy is the energy possessed by an object in motion. Kinetic energy is directly proportional to the object's mass and the square of its velocity:* K.E. = 1/2 m v2.

Comment: If there were no motion, not only would all objects stop, but they would disappear. Motion is vital to energy and matter; it is their life, their *l'énergie spirituelle*, to use Bergson's term. To put it simply, no motion, no mass. Motion equals mass, in a way. Concentration and "quantity" of motion from the micro to the macro level equals the quantity of mass. The bigger the total output of motion from the micro point to the measured point, the bigger the mass. The whole family of the Universe (Omniverse) is kept alive and together owing to the primordial Universal immaterial Source—Ultimate Force feeding motion and the whole of the Universe and existence with its underlying force—Ultimate Reality and the basis of Everything.

- *The Big Bang is either an action or reaction with a high enough escape velocity because, instantly, the Universe created a vast space at a much higher speed than the speed of light.*

Comment: This paradox can be explained by the Theory of the Absolute because there is no mass, not even the rest mass, so infinity is reached at the point of absolute speed, which translates into the infinity of everything. This point is the point of absolute zero. Only Zero can contain infinity. Before the Big Bang, zero is the "point" of absolute density and no space.

Im = mc
Im = 0
Av = 0

$0 = \infty$

$Im = Ag = 0 = \infty$

$\infty = 0$

Im (Infinite or absolute mass)

Av (absolute velocity)

Ag (absolute gravity)

Comment: Mass does not pull the fabric of space, but the motion of the "mass" (kinetic energy) pulls the fabric of the universe, without which space would lose sense. Space as an absolute void always stays the same and cannot be affected or pulled. As a result of the absolute pull, the world disappears. At the moment of the almost absolute speed, the pull itself becomes almost absolute, which is the point of almost absolute density inside a black hole or a point of singularity.

An increase in speed does not affect the size or volume of mass. Mass would not increase with the increase in speed. What happens is that with increasing speed, the "gravitational energy" or the "gravitational pull" increases. All this "additional" energy or force is the energy, force, or pull of the increased motion and its effect, not the increased mass that stays the same. At the speed of light, the mass, as a result of the speed of light, becomes infinite in terms of its "force" but not in terms of its volume, which stays the same. When the mass reaches the speed of light, or right before reaching the absolute speed, it becomes "infinite," which means that its "gravitational" force becomes almost "absolute," which equals the mass being infinite, not in reality but in effect. Right after the point of absolute speed (or the speed of light), the mass leaves the realm of the existing universe as we see it and disappears.

Primordial "kinetic energy" is the source of all physical forces and laws. Without motion, the laws of physics, the universe, and the spacetime continuum stop.

Even the rest mass is only mass as a consequence of existing laws of physics and existing motion functioning in a standard way.

In that sense, the disappearance of kinetic energy or motion would mean that there would be a disappearance of rest mass, too. Ultimately, all this means that the disappearance of motion simultaneously means the disappearance of energy.

Comment: Shortcuts for long travels throughout the universe are possible, although this does not imply time travel or a time machine. Time travel to the past or future would be possible if there were exact repetitions of previous "worlds" in the myriad of cosmic cycles and not by time travel *per se*.

Einstein—Theory of Relativity

- *There is no absolute space and time; the same for all observers.*
- *The velocity of light is limiting velocity. No object in the Universe can move at a speed greater than the speed of light.*
- *The conservation of energy and conservation of mass have been united into one law by the Theory of Relativity.*
- *Mass and energy have become interchangeable terms. As the body moves faster, its energy, and therefore its mass, increases. At the velocity of light, its mass would be infinite. Experiments confirmed Einstein's statements.*

De Broglie-Bohm theory

Louis de Broglie[44]:

Any particle, even isolated, has to be imagined as in continuous, 'energetic contact' with a hidden medium.

---

[44] Broglie, L. de. (Louis Victor Pierre Raymond, 7th Duc de Broglie; 1892—1987.) French physicist who made significant contributions to the field of quantum theory. He recognized the wave nature of electrons and that all matter has wave properties. The de Broglie hypothesis of wave–particle duality secured him the Nobel Prize in Physics in 1929.

- **Vis a tergo** – *A push from behind.* (**vis**—*force [the product of mass and acceleration]). We can identify* **vis a tergo** as *Aristotle's First Mover's move or push.*

Comment: Space, as an appearance (expression of an Ultimate Mind) and as we see it, is dimensional, but the Ultimate Mind is non-dimensional, for it is immaterial. It is infinitely finite and finitely infinite. Ultimate Mind (or Source) is de Broglie's "hidden medium." We can identify **vis a tergo** or a push from behind with de Broglie's "hidden medium" or Aristotle's First Mover.

Gravity

Def: *Gravitation, or gravity, is a natural phenomenon by which physical objects attract each other with force proportional to their masses. Gravitation is most familiar as the agent that gives weight to objects with mass and causes them to fall to the ground when dropped. Gravitation causes dispersed matter to coalesce and coalesced matter to remain intact, thus accounting for the existence of the Earth, the Sun, and most of the macroscopic objects in the Universe.*

Comment: Gravity is an effect of motion. We must distinguish between motion and gravity to understand both functions precisely. Instead of taking an abstract and somewhat obscure way, we state that there is no gravity without motion. We will build the argument and analyze the linguistic implications of the words (terms) representing different phenomena from this premise.

If gravity is more dependent on motion than mass, then the whole Universe, without motion, even if it existed as such, would not only be motionless but also a compilation of dead matter without weight (which would be the loss of mass), and, therefore, gravity. Such a Universe would be equal to nothingness.

- *Uncertainty in measuring on a microscopic level. We cannot know anything with absolute certainty.*

- *The only way to measure one particle and affect the other is for the signal traveling between them faster than the speed of light, which Einstein proved incorrect.*

(**Comment:** This may be correct based on the Theory of the Absolute and the functioning of the Universal Mind.)

The Dissolution of the Being

Before becoming (as a world) or transforming into the world, the Being is simultaneously absolutely finite and infinite (potentially). The Being must transform and escape its absolute nature to have a purpose and meaning. That's why it manifests itself through a living world.

Even though the Being accepts emptiness into itself, it is still, although predominantly empty, full of essence. Although everything may appear empty, no portion of "empty space" is immune and devoid of the laws of the Being (Ultimate Mind). The only space not affected by the living Being may be an absolute vacuum, space beyond space, or "absolute space" (if we can call it this way). On the other hand, although it exists or may exist, we can only treat absolute space theoretically because such space is not even space. After all, its infinity makes it finite, meaningless, and nonexistent. Every point in such a space is the same (because there is no point in such a space at all). That point is zero, and only zero possesses the potential for infinity, spatially and temporally.

Paradoxically, the Being creates space. Without the Being, there is no space except ultimate emptiness, which is meaningless without the creation of shapes in the emptiness. Once the Being envelopes nothingness and transforms in plurality, distances among shapes become possible. That is the creation of space as we know it. The dispersion of the Being into myriad shapes creates the Universe or the ultimate Omniverse. Space is not "spaceless" anymore; every spot becomes different since there are distances in the multitude of shapes, and the meaning is born. But the Being cannot materialize as the world (Universe) without nothingness.

JSWT Discoveries[45]

James Web Space Telescope discovered big galaxies and black holes when the Universe was only 500 to 700 million years old. Scientists previously thought it impossible for galaxies as big as the Milky Way to exist in the Universe's early stages. This will lead to new views about not only galaxies and black holes but also dark matter and energy and the overall picture of the Universe. Many old questions must be revisited, looked at with a fresh eye, and redefined.

[45] James Webb Space Telescope, the world's biggest telescope, was launched on December 25, 2021.

ABOUT THE AUTHOR

Dejan Stojanović (1959) was born in Peć, Kosovo (the former Yugoslavia). Although he is a lawyer by education, he has never practiced law and instead became a journalist, working as a foreign correspondent in the early nineties. He is also a poet, essayist, philosopher, and businessman.

He has published books of poems:
Circling (Krugovanje), Narodna knjiga—Alfa, Belgrade, has three editions: 1993, 1998, and 2000.
The Sun Watches Itself (Sunce sebe gleda), NIP Književna reč, Belgrade, 1999.
The Sign and its Children (Znak i njegova deca), Prosveta, Belgrade, 2000.
The Creator (Tvoritelj), Narodna knjiga, Belgrade, 2000.
The Shape (Oblik), Gramatik, Podgorica, 2000.
The Dance of Time (Ples vremena), Konras, Belgrade, 2007.
Pentalogy: *The World in Nowherness (Svet u nigdini):*
1. Ozar (Ozar), Udruženje književnika Srbije, Belgrade, 2017.
2. The World and God (Svet i Bog), Udruženje književnika Srbije, Belgrade, 2017.
3. The World in Nowhereness (Svet u nigdini), Udruženje književnika Srbije, 2017.
4. The World and Humans (Svet i ljudi), Udruženje književnika Srbije, Belgrade, 2017.
5. The Home of Light (Dom svetlosti). Udruženje književnika Srbije, Belgrade, 2017.
The Hidden Light (Skrivena svetlost), Čigoja, Belgrade, 2018.
Primordial Spark (Iskra iskona), Albatros plus, Belgrade, 2021.
Centuries and Steps (Vekovi i koraci), Albatros plus, Belgrade, 2023.

Essays:
Creator and Creating (Stvaralac i stvaranje), Albatros plus, Belgrade, 2021.
The New Man and the New World (Novočovek i novosvet), Rad, Belgrade, 2022.

Anthology: Selected Serbian Plays (Izabrane srpske drame), USA, 2016.

A book of his selected interviews, *Conversations*, was published in 1999 by *Književna reč*, Belgrade. The Serbian Heritage Foundation and the Association of Writers of Serbia awarded it the Rastko Petrović Prize for intellectual engagement. The book included interviews with several prominent American writers, including Nobel Laureate Saul Bellow, Charles Simic, and Steve Tesic.

In 1986, as a young writer, he was recognized among 200 writers at the Bor (former Yugoslavia) Literary Festival.

In addition to poetry and prose, he has worked as a journalist for the Serbian weekly magazine *Pogledi (Viewpoints).*

Made in United States
North Haven, CT
04 November 2026

21768655R00166